Internet of Things and BDS Application

Bo Wang · Xiangsheng Liu · Yaqi Zhang

Internet of Things and BDS Application

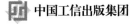  

Bo Wang
Beijing Institute of Technology
Beijing, China

Xiangsheng Liu
CNTEN Corp. Ltd.
Beijing, China

Yaqi Zhang
Mianyang TV Station
Mianyang, Sichuan, China

ISBN 978-981-16-9196-6   ISBN 978-981-16-9194-2   (eBook)
https://doi.org/10.1007/978-981-16-9194-2

Jointly published with Publishing House of Electronics Industry
The print edition is not for sale in China (Mainland). Customers from China (Mainland) please order the print book from: Publishing House of Electronics Industry.

This Springer imprint is published by the registered company Springer Nature Singapore Pte Ltd.
The registered company address is: 152 Beach Road, #21-01/04 Gateway East, Singapore 189721, Singapore

Preface

With the continuous development of the new generation of information technology, the combination of the Internet of things (IoT), which provides perceptual information, and the satellite navigation system, which is the spatiotemporal benchmark, is becoming closer and closer. As China's independent and controllable spatiotemporal benchmark, Beidou satellite navigation system (BDS) provides complete global services from 2020. The key of technology lies in the application. The close combination of precise spatiotemporal information with the interconnection of all things not only empowers a variety of industries, but also leads the reform of more industries. According to the statistics of GNSS and LBS Association of China (GLAC), the total output value of Chinese satellite navigation and location-based services industry has exceeded 400 billion Yuan in 2020, of which BDS has contributed more than 80% to the core output value of the industry. Therefore, the continuous development of IoT and BDS application is not only related to the national security of China, but also greatly promotes economic development of China. With the development of 5G, cloud computing, big data, artificial intelligence, and a series of emerging technologies, the IoT and BDS application will bring more vigorous vitality to the development of the industry.

BDS is a global satellite navigation system built and operated independently by China and compatible with other satellite navigation systems in the world. Since the 1990s, China has completed the construction of BDS-1, BDS-2, and BDS-3 systems in accordance with the "three-step" strategy, "First active and then passive, first regional and then global," and has blazed a road of satellite navigation system construction with Chinese characteristics.

Guided by the BDS spirit of "Independent innovation, unity and cooperation, overcoming difficulties and pursuing excellence," BDS system provides global service capability through high-density networking, ensures stable service with high operation speed, creates new highlights of international cooperation with high standards, and drives new breakthroughs in application with high precision. The BDS system has reached the level of 10 m at 95% confidence level, the positioning accuracy of 10 m height, the speed of 0.2 m/s, the timing accuracy of 20 ns, and the availability of system services more than 95%. In the Asia-Pacific region, the positioning accuracy of 5 m is achieved at 95% confidence level. People all over the world can enjoy the service of BDS system. Through the BDS ground-based augmentation system, the precision of the spatiotemporal information provided by BDS can be improved to the centimeter level of real time and millimeter level of post-processing. The accurate spatiotemporal information can be widely used in various industries with extremely low cost and convenient way.

IoT is a network based on perception, which achieves the full interconnection of people and things. At present, the IoT industry has developed from the stage of technology improvement to the stage of application popularization. With the continuous improvement of all kinds of near-field communication technology and low-power communication technology, artificial intelligence technology continues to emerge. Under the dual promotion of supply-side reform and demand-side reform, the IoT industry has set off the third development wave represented by basic industries and scale consumption. Relevant data show that the scale of the global IoT industry has increased from $50 billion in 2008 to $151 billion in 2018. By 2022, IoT market of China is expected to reach $255.23 billion, accounting for 24.3% of the global total expenditure in the same period, after 25.2% of the USA, ranking second in the world, and the vision of IoT is being realized.

The key of technology development lies in application. Only the stage of popularization of application is the stage of value creation of various technologies. Driven by technologies such as precise spatiotemporal information and interconnection of all things, popularization of industrial applications has become the top priority of technology development. Technology not only empowers the industry, but also leads the change of the industry.

Under the new situation, the industry application of BDS should change from focusing on platform construction to focusing on platform capability, deeply combining the precise spatiotemporal information provided by the ground-based augmentation system with the industry application, improving the platform capability and enabling the industry, and changing from focusing on scale and quantity to focusing on quality and connotation. The scale effect of the BDS has been highlighted. More importantly, the application quality must be upgraded to a certain level. From the traditional division to the integration of innovation, the development of innovative applications is carried out in line with the traditional application fields such as transportation, logistics, safety and emergency, and other industries. The business information provided by IoT sensors is closely combined, so that the application of BDS is no longer limited to positioning, navigation, and timing. Instead, it takes spatiotemporal information as the starting point, perception as the basis, and information transmission as the link to improve the application level of the industry in an all-round way. It changes from focusing on advantages to focusing on sharing, sharing the advantages formed in a certain industry with more industries, and sharing BDS spatiotemporal information to more and more industries. We continue to practice the concept of "Good in the sky, good on the ground" by academician Sun Jiadong, the first chief designer of BDS.

Dr. Wang Chenglong, Dr. Jia Jingyuan, Dr. Ma Zixuan, Master Liu Jingyang, and Master Chen Jiaqi gave great efforts in the process of compiling this book.

The industry application analysis contents in this book are supported by GNSS and LBS Association of China and its precision application professional committee and Shenzhen Beidou Industrial Internet Institute. The publication of this book is also supported by Beijing Science and Technology Plan (Z181100003518003) and Beijing Jinqiao Project Seed Fund (ZZ19018). Chapters 1–5, 7, Sects. 6.2–6.4, and 6.7–6.9 are written by Bo Wang, Sects. 6.1, 6.5, and 6.6 are written by Xiangsheng Liu, and the whole book is translated by Yaqi Zhang. After the completion of the first draft of this book, Dr. Miao Qianjun, Prof. Cao Chong, and Prof. Zhang Quande reviewed it and put forward very valuable opinions and suggestions. We would like to express special thanks to them.

IoT and BDS, as developing high tech, are not only involving the intersection of various disciplines, but also highly engineering application technology. Limited to the author's knowledge level and the scope of data access, omissions in the book are inevitable. It would be grateful if readers criticize and correct them.

Beijing, China   Bo Wang
2021   Xiangsheng Liu
                                                                                            Yaqi Zhang

Contents

Chapter 1
Global Navigation Satellite System

1.1 Overview of Global Navigation Satellite System

Global navigation satellite system (GNSS) is a navigation system that uses the signals from navigation satellites to determine the position of the carrier. In 1957, the first man-made satellite successfully launched by the Soviet Union became the cornerstone of the development of GNSS. People were no longer confined to the ground, but began to look into the space. With the development of space science and technology, traditional technology also radiated new vitality. Two researchers from Johns Hopkins University in the United States, by observing the radio signals transmitted by satellites, firstly connected the Doppler frequency shift of common signals on the ground with the trajectory of satellites. They put forward a method to obtain the precise position of satellites in space by measuring the Doppler frequency shift of signals broadcasted by satellites using ground observation stations with known positions. The effectiveness of this method was verified by the joint observation experiment of the ground observation station to the satellite, which determined the satellite trajectory. Based on the experiment results, two other researchers from the same university put forward another resolution, which is, if the precise position of the satellite is known, the precise position of the ground observation station can be determined by measuring the Doppler frequency shift of the satellite signal on the ground. Thus, the basic concept of satellite navigation and positioning came into being.

Satellite navigation system began to develop in the mid-1960s, and now it has been widely used in various fields of civil and military, bringing huge economic and social benefits. It is precisely because satellite navigation system plays a vital role in many applications that lots of countries are trying to build their own independent satellite navigation system.

GPS in the United States is the most widely used satellite navigation system at present. Its application scope has covered many industries in the world. GLONASS in Russia is greatly behind GPS due to various factors. But after more than ten years of satellite network replenishment, the system reconstruction has been completed.

© Publishing House of Electronics Industry 2022
B. Wang et al., *Internet of Things and BDS Application*,
https://doi.org/10.1007/978-981-16-9194-2_1

Galileo system planned by the European Union has also basically completed the launch of networking satellites, and started the service. China has also completed the construction of BDS-3 on the basis of BDS-1 and BDS-2. BDS-3 was fully completed and started to provide global services in 2020. At present, all countries are promoting the compatibility operation of GNSS and striving to achieve benign interaction and cooperation.

1.1.1 BDS

Beidou satellite navigation system (BDS) is a global satellite navigation system independently developed and operated by China. Together with GPS of the United States, GLONASS of Russia and Galileo of the European Union, BDS is known as the world's four major satellite navigation systems. BDS started the official service of regional navigation, positioning and timing on December 27, 2012. Beidou-2 system, composed of 16 navigation satellites, serves most of the Asia Pacific region, including China and its surrounding countries and regions. In July 31, 2020, Beidou-3 was fully completed with 30 satellites.

1.   Composition of BDS

In the 1980s, with the completion of the GPS, China put forward the idea of establishing an autonomous and controllable temporal-spatial reference satellite navigation system, and completed the construction of BDS-1 in 2003 as the experimental system. In 2004, China began to prepare to build the Beidou global satellite navigation system on the basis of the Beidou experimental system. Compared with other global satellite navigation systems (such as GPS, GLONASS, etc.), BDS not only has the functions of navigation, positioning and time service, but also has a unique communication function, namely short message communication.

   BDS is a new generation satellite navigation system which is built, maintained and operated by China. Different from the experimental system BDS-1, which has active positioning and timing mode, BDS-2 and BDS-3 adopts passive positioning method and broadcast satellite navigation signal. This method overcomes many shortcomings of BDS-1, such as active positioning, area coverage and poor system survivability. The completed BDS-3 is similar to GPS and GLONASS in principle, and is also a global satellite navigation system.

   The construction and development strategy of BDS is divided into three stages.

(1)   In the first stage, BDS-1 was built. Since 2000, China has successfully launched three Beidou satellites in three years, and built the Beidou experimental system with dual satellite positioning structure, including two working satellites and one backup satellite. The system can provide basic positioning, timing and short message services with active mode. Users need to send positioning request signals to satellites, which is inconvenient to use and has poor survivability.

(2)   In the second stage, the regional system BDS-2 was constructed. In April 2007, Compass-M1, the first MEO satellite of BDS-2 regional system, was successfully launched. This satellite ensured orbit and frequency resources, and completed a large number of technical experiments. On April 15, 2009, the first GEO (geostationary orbit) satellite (Compass-G2) of BDS-2 regional system was successfully launched by Long March-3C carrier rocket. This satellite verified the related technology of GEO navigation satellite. In October 2012, BDS-2 regional system successfully launched the 16th satellite, and completed the satellite networking. BDS-2 was officially put into operation on December 28, 2012, providing navigation, passive location, timing and other services for the Asia Pacific region.

(3)   (3) In the third stage, the global system BDS-3 was constructed. On July 31, 2020, BDS-3 completed the global coverage and provide high accuracy and high reliability for timing, positioning and navigation services.

The basic principles of BDS construction are openness, autonomy, compatibility and gradualness. Openness refers to BDS opening to the world and providing free and high-quality services. Autonomy refers to the independent development and operation of the BDS by China. Compatibility refers to the compatibility and interoperability with other satellite navigation systems. Gradualness refers to the progress of BDS combining of economic and technological development of China. BDS is with the progressive mode of development, to ensure the smooth transition of the system construction phase, and ultimately provide users with continuous long-term global services.

BDS plans to provide users with global services and regional services. The global services include free open service with positioning accuracy of 10 m, timing accuracy of 50 ns and speed measurement accuracy of 0.2 m/s, and authorized service with higher accuracy and reliability under complex conditions. The regional services include wide area differential service with positioning accuracy of 1 m, and short message communication service, which has been expanded to global service in BDS-3.

BDS can be divided into space segment, control segment and user segment.

(1)   The space segment consists of three geostationary orbit (GEO) satellites and 27 non geostationary orbit (Non-GEO) satellites. The specific layout of the complete constellation configuration is "GEO + MEO + IGSO". The constellation scheme of BDS-2 regional navigation system consisted 5 GEO satellites, 3 IGSO satellites and 4 MEO satellites. Five GEO satellites are fixed at relatively stationary points above the earth's equator. Four MEO satellites are running on the orbit with radius of 21,500 km. Three IGSO satellites are on different orbit surfaces with the radius of 36,000 km. BDS-3 global navigation system consists of three GEO satellites, three IGSO satellites and 24 MEO satellites, with 30 satellites in total. Such a constellation design ensures that signals from more than four navigation satellites can be received at any time at any point on the earth, and even signals from more than ten navigation satellites can be received in areas with good observation conditions.

(2)   The control segment includes a series of ground stations such as monitoring station, time synchronization uplink station and master control station, as well as inter-satellite link operation and management facilities. The main control station is the core of the ground segment as well as the whole satellite navigation system. It has the functions of monitoring satellite constellation, maintaining time benchmark, updating navigation message and so on. The function of the time synchronization uplink station is to inject the information sent from the master control station into each satellite. These information include navigation messages, wide area differential information and time synchronization signals. The function of the monitoring station is to monitor the satellite and complete data acquisition. The monitoring station continuously observes the satellite constellation to form the monitoring data. Then it collects the satellite, meteorological and other information and transmits to the main control station for processing.

(3)   The common components of user segment include positioning chip in mobile phone, handheld receiver, vehicle receiver, navigation and aerospace application receiver, etc. The user terminal is the equipment which can solve the position, velocity, time (PVT) in the whole satellite positioning system. According to the design principle of compatibility, the user terminal of BDS will be able to achieve good compatibility and interoperability with other GNSS such as GPS, GLONASS and Galileo. At present, the user terminal of BDS has been widely used in the market. Relevant policies and standards are being formulated.

2.   Time system and coordinate system

1)   Time system

The principle of positioning by satellite navigation system is the basic formula of "distance = speed × time", that is, measuring time first and then converting to distance. For example, GPS system defines a special time system GPST (Global Positioning System Time) based on atomic time, which has an integer second relationship with UTC. Since January 1, 2006, GPST = UTC + 14, and the non-integer second error between GPST and UTC is usually controlled within 40 ns.

   Because the propagation speed of satellite signal is the speed of light, accurate time system is very important. BDS defines another time system, which is called BDT (Beidou Time). The starting epoch time of BDT is 00:00 on January 1, 2006 of UTC, and the deviation between BDT and UTC is less than 100 ns.

   The design of BDS is based on the principle of compatibility. In order to guarantee the compatibility and interoperability with other systems, the interoperability between BDT, GPST and Galileo is also considered at the beginning of design. The time difference between BDT and other time systems will be monitored and broadcast.

2)   Coordinate system

The coordinate system currently used by GPS is the World Geodetic System (WGS-84) proposed by the Defense Mapping Agency (DMA) in 1984. The speed and position of satellite calculated by GPS receiver are directly expressed in WGS-84.

BDS uses China Geodetic Coordinate System 2000 (CGCS2000 for short). The coordinate system has been used since 2008, and the transition period is 8–10 years. The coordinate system is not only compatible with BDS, but also highly consistent with the International Terrestrial Reference Frame (ITRF). The difference is about 5 cm, therefore, it is very convenient to use. For most applications, the coordinate conversion between CGCS2000 and ITRF could be ignored.

3. Signal characteristics

Beidou satellite signal consists of three parts: navigation message (data code), pseudo-random noise code (divided into authorized and open services) and carrier. Users need to interpret the satellite navigation message from the satellite signal, and then calculate the current real-time position of the satellite through a series of algorithms. On one hand, the pseudo-random noise code is used to modulate the data code. On the other hand, it is used to distinguish the source of the received satellite signal. The pseudo-random noise code which provides different services also encrypts the satellite signal to distinguish different authorized users. Using pseudo-random noise code and carrier, the distance between satellite and receiver can be measured. Then the position and velocity information of user can be calculated by using the satellite position calculated from navigation message.

Because the satellite runs in the space tens of thousands of kilometers away from the earth, in order to make the satellite signal propagate to the ground through the atmospheric ionosphere and troposphere, it must be transmitted in the Ultra High Frequency (UHF) band. The data code modulated by pseudo-random code is modulated to the carrier, to achieve the purpose of long-distance signal transmission.

There are three carrier frequency bands applied by BDS. Signals will be transmitted in different ways in B1, B2 and B3 frequency bands. The specific frequency range is B1: 1559.052–1591.788 MHz; B2: 1166.22–1217.37 MHz; B3: 1250.618–1286.423 MHz.

According to the gradual principle of the construction of BDS, the composition of the BDS constellation and the broadcast signal transmission stage will be different in different construction stages, as shown in Table 1.1.

There are great changes between BDS-2 regional system and BDS-3 system in many aspects, such as center frequency point, code transmission rate, modulation mode, use right and so on.

Table 1.1 Different stages of the BDS constellation and the corresponding broadcast signal transmission

Year	Constellation	Signals
2012	5 GEO + 3 IGSO + 4 MEO (regional service)	The second phase signal
2020	3 GEO + 3 IGSO + 24 MEO (global service)	The third stage signal

Table 1.2 Signal characteristics of BDS-2 regional system

Signals	Center frequency point (MHz)	Code rate (cps*)	Bandwidth (MHz)	Modulation mode	Service type
B1(I)	1561.098	2.046	4.092	QPSK	Open
B1(Q)		2.046			Authorized
B2(I)	1207.14	2.046	24	QPSK	Open
B2(Q)		10.23			Authorized
B3	1268.52	10.23	24	QPSK	Authorized

*cps, short for chips per second, is a chip rate unit

Table 1.3 Characteristics of BDS-3 satellite signals

Signals	Center frequency point(MHz)	Code rate(cps)	Modulation mode	Service type
B1-CD	1575.42	1.023	MBOC(6, 1, 1/11)	Open
B1-CP		2.046	BOC(14, 2)	Authorized
B1-A				
B2aD	1191.795	10.23	AltBOC(15, 10)	Open
B2aP				
B2bD				
B2bP				
B3	1268.52	10.23	QPSK(10)	Authorized
B3-AD		2.5575	BOC(15, 2.5)	Authorized
B3-AP				

Table 1.2 shows the signal characteristics of BDS-2 regional system. In this stage, it focuses on regional navigation and positioning functions, and provides two kinds of positioning services with different permissions for the Asia Pacific region.

Table 1.3 shows the characteristics of BDS-3 satellite signal. Compared with BDS-2 regional system signal, only the center frequency point and modulation mode of B3 band have not changed, while the center frequency point and modulation mode of other bands have changed. The modulation mode has changed from QPSK to BOC.

Compared with BDS-2, BDS-3 adds a variety of satellite signals, provides more modulation modes, and adds signal types in the two kinds of service.

1.1.2 GPS

GPS (Global Positioning System) is established by the U.S. Department of defense to meet the requirements of military usage for high-precision navigation and positioning. GPS can provide real-time, all-weather and global navigation services for the army of land, sea and air. GPS also can carry out some military activities such as intelligence collection, nuclear explosion monitoring and emergency communication. The system was founded in 1973. After more than 20 years of project demonstration, engineering development and launch networking, 24 GPS satellite constellations with 98% global coverage were built in 1994. The global services were provided in 1995.

As a new generation of satellite navigation system developed after the U.S. Transit satellite navigation system, GPS provides three-dimensional navigation and positioning capabilities with the advantages of global coverage, all day, all-weather and continuity. As an advanced measurement, positioning, navigation and timing system, GPS has been integrated into all aspects of national security, economic construction and people's livelihood development.

1.   Composition of GPS

GPS consists of three parts: space segment, control segment and user segment.

1)   Space segment

The space segment is composed of 21 working GPS satellites and 3 standby satellites on orbit, forming a complete "21 + 3" GPS satellite working constellation. GPS has six orbital planes, numbered as A, B, C, D, E and F, respectively. Four satellites are evenly distributed on each orbital plane. The inclination of the orbital plane relative to the equatorial plane is 55° and the angle between the orbital planes is 60°. This constellation configuration can ensure that at least four satellites with good geometric relationship can be observed for positioning at any place and any time on the earth. The average orbit height of GPS satellite is 20200 km, and it runs in a near circular orbit every 11 h and 59 min (star time).

2)   Control segment

The control segment of GPS is composed of one main control station, three injection stations and several monitoring stations distributed all over the world. The main control station is located at the Consolidated Satellite Operations Center (CSOC) in Colorado Springs, USA. Its function is to receive the observation data of GPS satellites from monitoring stations all over the world. It calculates the correction parameters of the ephemeris and clock of GPS satellites, and edits these data into navigation messages. The navigation message generated by the master control station is sent to the GPS satellite in the form of S-band through the injection station. Then the GPS satellite will broadcast the navigation message modulated by carrier and ranging to the user in real time. The main control station is also responsible for controlling the GPS satellite. For example, when the working satellite fails, the main control station is responsible for dispatching the standby satellite to replace the faulty satellite.

3)   User segment

The space segment and control segment of GPS are used as infrastructure to provide navigation, positioning and timing services for military and civil users, which are widely used in various fields. Through GPS receiver, users can receive and calculate satellite signals to obtain navigation, positioning and timing services. GPS receiver receives satellite signal through antenna, converts signal by RF front-end, processes observation data by baseband, and obtains navigation, positioning and time information by navigation algorithm.

2. GPS satellite signal

The traditional GPS satellite will transmit two kinds of carrier signals, namely L1 and L2 carriers. The frequencies of the two carriers are 1575.42 and 1227.60 MHz, respectively. Their wavelengths are 19.03 and 24.42 cm, respectively. The ranging code and navigation message are modulated on L1 and L2 carriers.

C/A code: Coarse/Acquisition Code is modulated on the L1 carrier. It is a 1 MHz pseudo random noise (PRN) code with a code length of 1023 bits and a period of 1 ms. It is the main signal used by ordinary civil users to measure the distance between receiver and satellite.

P Code: Precision Code is modulated on L1 and L2 carriers. It is a 10 MHz pseudo-random noise code with a period of 7 days. It can only be used by US military users or authorized users.

Navigation message: the navigation message contains the orbit parameters of GPS satellite, satellite clock correction parameters, satellite almanac and some other system parameters. It is modulated on L1 carrier and its signal frequency is 50 Hz. Through the ephemeris parameters in the navigation message, the user can calculate the instantaneous position of the GPS satellite in orbit, and also calculate the time through the clock correction.

There are many kinds of signals broadcast by GPS. In practical navigation and positioning applications, one or more signals can be processed at the same time. Generally, L1 and L2 carrier phase observations, C/A code and P code pseudo range modulated on L1 and L2 carrier, and Doppler frequency shift of L1 and L2 carrier are used. For different application requirements, in addition to carrier and PRN codes, different combinations of the above observations are also used, such as single difference, double difference and triple difference observations of carrier phase, wide lane observations and narrow lane observations.

With the continuous advancement of GPS modernization plan, the number of civil GPS signals has increased from the original one L1 C/A code signal to four. In addition to L1 C/A code, civil GPS signals loaded on L2C, L5 and L1C carriers are also provided. The frequency point of L2C signal is 1227.6 MHz, which aims to correct the ionosphere delay of civil dual frequency receiver. Due to the higher effective power of L2C signal, it can achieve signal acquisition faster and improve the reliability of navigation and positioning. The frequency point of L5 signal is 1176.45 MHz, which is the protection band of radio for aviation safety service. By combining with L1 C/A code signal, it can provide dual frequency ionosphere correction for airborne GPS receiver, with improvement of the accuracy and reliability of navigation and positioning. The frequency of L1C signal is 1575.42 MHz, which is designed for interoperability with Galileo system and other GNSS. It can be backward compatible with the existing L1 signal. L1C signal can solve the problem that civil signal is easily blocked, and improve navigation and positioning effect of the receiver.

3. Advantages and disadvantages of GPS

1) Advantages of GPS

The basic principle of GPS is "time measurement to distance measurement". By measuring the signal propagation time the GPS receiver can obtain the signal propagation distance, and then carry out positioning and navigation. The system takes the high-precision atomic clock as the core, and provides a large range of passive positioning and navigation by broadcasting specific signals. Therefore, GPS has the following advantages.

(1) Global coverage. There are 24 satellites in the space segment of GPS. Satellites in the constellation are evenly distributed, and the orbit is as high as 20200 km. Therefore, more than 4 satellites can be synchronously observed at any point on the earth and near space, providing global and all-weather continuous navigation and positioning service.

(2) High accuracy 3D positioning. GPS can continuously provide three-dimensional position, three-dimensional velocity and accurate time information for land, sea, air and space users. PRN code can achieve 5–10 m single point positioning accuracy. Pseudo range difference and carrier phase difference can achieve sub-meter level, centimeter level and even millimeter level positioning accuracy, which can meet the accuracy requirements of different applications.

(3) Passive navigation and positioning. GPS satellites are constantly broadcasting signals, therefore, the receiver only needs to passively receive satellites signals to output navigation and positioning information, instead of transmitting any signals to the outside. Passive navigation and positioning not only has good concealment, but also can accommodate unlimited users theoretically.

(4) Real time navigation and positioning. GPS receivercan output real-time positioning data with update rate of 1–100 Hz, which can meet the high dynamic needs.

(5) Good anti-jamming performance and strong confidentiality. GPS applies code division multiple access (CDMA) technology and uses different pseudo-random noise codes to distinguish different satellites. In particular, the P-code has good anti-interference and confidentiality due to its higher power, longer code length and better security measures.

2) Shortcomings of GPS and its improvement

The advantages of GPS are obvious, but there are also some shortcomings.

(1)   No communication link. GPS is a passive navigation and positioning system.
       There is no satellite communication link between different users. Therefore, it
       cannot meet the needs in some special occasions, such as emergency rescue,
       aviation control, location report and etc.. In practical application, the solutions
       of GPS combined with satellite communication or mobile communication are
       always applied.
(2)   The signal is easily blocked. Due to the limitation of satellite signal broad-
       casting power, GPS satellite signal is very weak from space to ground receiver.
       It is easy to be blocked by tall buildings, trees and so on, resulting in the decline
       of navigation and positioning accuracy. In practical application, GPS and INS
       are usually integrated. In addition, the GPS modernization program is proceed
       to develop new GPS satellites and enhance the signal power.
(3)   The signal has no water entry capacity. GPS signal is L-band and has no ability
       to enter water. Therefore, all kinds of submersibles must float upward on surface
       to use GPS, or release floating antenna to the surface. In order to solve this
       problem, GPS/INS or GPS/radio integrated navigation system can be applied.

1.1.3 GLONASS

GLONASS (Global Navigation Satellite System) is a global navigation satellite
system constructed by the Soviet Union. The first satellite was launched on October
12, 1982. The design of all satellites (24) was completed and the whole system
began to operate on January 18, 1996. With the disintegration of the Soviet Union,
GLONASS is currently managed and maintained by the Russian space agency.
Similar to GPS, GLONASS can also provide global real-time, all-weather three-
dimensional continuous navigation, positioning and timing services for civil and
military users on land, sea, air and space.

1.   GLONASS composition

Similar to GPS, GLONASS is also composed of three parts: space segment, ground
segment and user segment. But the specific technology of each part is quite different
from GPS.

1)   Space segment

The constellation of GLONASS is also made up of 24 GLONASS satellites, including
21 working satellites and 3 backup satellites. With Russia continuous maintenance
of GLONASS, the 21 satellites currently forming the constellation are GLONASS-
M or GLONASS-K. The 24 satellites are evenly distributed on three orbital planes.
The three orbital planes form an angle of 120° with an inclination of 64.8° and an
orbit height of 19,100 km. The orbit eccentricity is 0.01 and the operation period is
11 h and 15 min. Eight satellites are evenly distributed on each orbit. Since the orbit
inclination of GLONASS satellite is greater than that of GPS satellite, the visibility of
GLONASS satellite is better in high latitudes above 50 degrees. Because GLONASS

Table 1.4 Constellation distribution of GLONASS

Track number		Satellite distribution ("–" means no satellite)							
Track 1	Satellite	1	2	3	4	5	6	7	8
	Frequency number	01	−4	05	06	01	−	05	06
Track 2	Satellite	9	10	11	12	13	14	15	16
	Frequency number	−2	−7	00	−	−2	−7	00	−
Track 3	Satellite	17	18	19	20	21	22	23	24
	Frequency number	04	−3	03	02	04	−3	03	02

mainly considers that Russian land area is large in high latitudes when designing and constructing, in order to ensure full coverage, its satellite orbit must be different from six orbital planes of GPS. GLONASS constellation in the complete case, can ensure that anywhere on the earth, at any time can receive at least four satellite signals. Therefore, users can obtain reliable navigation and positioning information. The distribution of specific satellite constellations is shown in Table 1.4.

Cesium atomic clock is available on each GLONASS satellite to generate high stable time and frequency standards, and provide high stable synchronization signals to all onboard computers. The onboard computer processes the information uploaded by the ground control part, generates navigation messages, ranging codes and carriers to broadcast to users. The information transmitted to the satellite by the ground control part is used to control the satellite operation in space. Navigation messages include the ephemeris parameters of the satellite, the offset value of the satellite clock relative to GLONASS UTC, the satellite health status and GLONASS satellite calendar, etc. Like GPS, GLONASS can transmit both civil and military codes.

2) Ground segment

The GLONASS ground segment maintains and controls the GLONASS constellation and satellite signals as a whole. It includes a system control center and a network of tracking control stations scattered throughout Russian territory. The ground segment is responsible for tracking and processing the orbit and signal information of GLONASS satellites, and transmitting control information and navigation messages to each satellite. After the collapse of the Soviet Union, GLONASS was managed by the Russian space agency. The ground segment was reduced to only sites in Russia. The ground segment consists of six components: system control center (SCC), telemetry and tracking command station (TT&C), uplink station (ULS), monitoring station (MS), central clock (CC) and satellite laser ranging station (SLR).

The function of the ground segment mainly includes the following aspects: (1) Measurement and prediction of the ephemeris of each satellite; (2) Tracking, control and management of the satellites; (3) Uploading the predicted ephemeris, clock correction value and almanac information into each satellite to generate navigation messages; (4) Synchronization of each satellite clock with GLONASS time; (5) Calculation of the deviation between GLONASS time and UTC; (6) Monitoring GLONASS navigation signal.

3)   User segment

The user equipment (receiver) of GLONASS can receive the navigation signal transmitted by satellites, including pseudo-random noise code and carrier phase. The
pseudo range and pseudo range rate can be calculated. The navigation message from
the satellite signal can be extracted and processed. By processing the navigation
message and pseudo range information, the position, speed and time information of
the user can be calculated.

The development of GLONASS user equipment is relatively slow. In addition to
historical reasons such as imperfect GLONASS constellation and unstable system
operation, GLONASS uses frequency division multiple access technology, which
makes user equipment more complex. As a result, the development and production
cost of GLONASS receiver is high. There are few types of receivers, limited functions, and high power consumption. However, as a global satellite navigation system
with the same development period and function as GPS, its application potential
is constantly highlighted with Russian continuous improvement. Since GLONASS
and GPS are the same or similar in system composition, working frequency band,
positioning principle, ephemeris data structure and signal debugging mode, GPS
and GLONASS receivers can be combined to receive satellite signals together. At
present, GPS/GLONASS Integrated receiver has also been widely used. The redundancy brought by multiple satellite navigation system improves the reliability of the
previous receiver using a single system.

GPS/GLONASS Integrated receiver can receive and process the signals of the
two systems at the same time, which has the following advantages. Users can receive
GPS and GLONASS satellite signals at the same time. The increase of the number of
observation satellites will significantly improve the geometric distribution of observation satellites and improve the positioning accuracy. Due to the increase of the
number of insight satellites, in some areas that are easy to block the signal, such as
cities, valleys and forests, users can obtain more accurate measurement, navigation
and monitor. In addition, the use of two independent satellite positioning systems
for navigation and positioning measurement can check each other, bringing higher
reliability and security.

2.   GLONASS time system

GLONASS time is the time reference of the whole navigation system. It belongs to
the UTC time system. Different from GPS, GLONASS takes UTC (Su) maintained by
Russia as the time reference. The difference between UTC (Su) and the international
standard UTC maintained by the Bureau International des Poids et Mesures (BIPM)
is within 1 μs. There is an integer difference of 3 h between GLONASS time and
UTC (Su), and the difference in seconds is less than 1 ms. The navigation messages
broadcasted by GLONASS satellites contain the parameters related to UTC (Su)
when GLONASS is used.

UTC is a time system based on atomic time seconds, which is as close as possible
to universal time. Due to the influence of the earth's pole shift and the non-uniformity
of its rotation, there is a difference between the two, and the difference expands with

time. In order to ensure that the difference between the two is not too large, UTC has a second jump, also known as leap second. Because GLONASS time belongs to UTC, there are also leap seconds. GLONASS time is corrected by leap second according to the notice of BIPM. Because there is no leap second in GPS, the difference of time between GPS and GLONASS needs to be considered when they are combined.

GLONASS time system includes two time scales: GLONASS system time and GLONASS satellite time. GLONASS system time is generated by the central synchronizer in ground segment, which is the time of the whole system. GLONASS satellite time is generated by the atomic clock equipped on each satellite, which is a kind of atomic time. Due to the existence of leap second, GLONASS system time and GLONASS satellite time are not exactly the same. The correction of GLONASS satellite time relative to GLONASS system time and UTC (Su) is calculated at GLONASS ground integrated control station, and uploaded to the satellite once every two days.

3.   GLONASS coordinate system

GLONASS applied the Soviet Union 1985 geocentric coordinate system (SGS-85) before 1993, and PZ-90 after 1993. PZ-90 belongs to the Earth Centered Earth Fixed (ECEF) coordinate system. According to official information, GLONASS coordinate system was updated from PZ-90 to PZ-90.02 at the end of 2006. The difference between GLONASS and ITRF remains at decimeter level. Between PZ-90.02 and ITRF2000, there is only the translation of the origin, which is -36 cm, $+8$ cm, and $+18$ cm in the direction of X, Y, Z, respectively.

The definition of PZ-90 coordinate system in ICD released by GLONASS is as follows. (1) Coordinate origin is located at the mass center of the earth; (2) Z-axis points to the protocol polar origin recommended by IERS (the average North pole from 1900 to 1905). (3) X-axis points to the intersection of the earth's equator and the zero meridian defined by BIH. (4) Y-axis satisfies the right-handed coordinate system. Because there are inevitably coordinate errors and measurement errors in the location coordinates of the tracking station, there are some differences between the defined coordinate system and the actual coordinate system. The reference ellipsoid parameters and other parameters used in PZ-90 coordinate system are shown in Table 1.5.

4.   GLONASS satellite signal

Similar to GPS, GLONASS also transmits L1 and L2 carrier signals in L-band. The L1 signal of GLONASS is modulated with P code, C/A code and navigation message, while the L2 signal is modulated with P code and navigation message. C/A code is used to provide standard positioning for civil organizations, while P code is used for high-precision positioning of Russian military or some authorized users. After 2005, by the request of ITU, Russia has transferred the L1 carrier frequency of GLONASS to 1598.0625–1606.5 MHz and the L2 carrier frequency to 1242.9376–1249.6 MHz.

GLONASS-M satellite is the second generation of GLONASS navigation satellite. Compared with the first generation, it has the advantages of high precision and long service life. It adds L2 civil code and can send more navigation messages. It

Table 1.5 Reference ellipsoid parameters and other parameters used in PZ-90 coordinate system

Angular velocity of earth rotation	$7,292,115 \times 10^{-11}$ rad/s
Gravitational constant of the earth (GM)	$398,600.44$ km^3/s^2
Atmospheric gravitational constant (fMa)	0.35×10^9 m^3/s^2
Speed of light (c)	$299,792,458$ m/s
Long radius of reference ellipsoid (a)	$6,378,136$ m
Oblateness of reference ellipsoid (f)	1/298.257839303
Acceleration of gravity (equator)	978,032.8 mgal
Correction value of gravity acceleration caused by atmosphere (sea level)	-0.9 mgal
Second order band harmonic coefficient of spherical harmonic function of gravitational potential (J_2)	$108,262.57 \times 10^{-8}$
Fourth order band harmonic coefficient of spherical harmonic function of gravitational potential (J_4)	-0.23709×10^{-5}
Reference ellipsoid normal gravity potential (u)	$62,636,861.074$ m^2/s^2

greatly improves the performance of GLONASS. GLONASS-K satellite is the third generation of GLONASS navigation satellite. Its service life will be extended to 10–12 years, and the third civil L3 band will be added.

GLONASS applied frequency division multiple access (FDMA) to identify different satellites, that is, the carrier frequency of navigation signals broadcast by each satellite is different. However, the frequency interval between adjacent satellites is the same. The frequency interval of L1 carrier is 0.5625 MHz and that of L2 carrier is 0.4735 MHz. Due to the FDMA mode, there will be more frequency points, which will occupy a wider frequency band. The L1 channel of 24 GLONASS satellites needs to occupy a bandwidth of about 14 MHz. Due to the limited space frequency resources, at the request of ITU, Russia stipulates that two GLONASS satellites with opposite positions on the same orbital plane use the same carrier frequency. This reduces the number of satellite carrier frequency channels to 12, meeting the requirements of reducing bandwidth and frequency channels.

5.   Comparison between GLONASS and GPS

GLONASS and GPS are the two earliest GNSS. They have similarities and differences in system composition, signal structure, signal type, coordinate system and time system. The similarities between the two systems mainly include the following aspects.

(1)   In terms of system composition, the two systems are composed of space segment, ground segment and user segment.
(2)   In satellite constellation, the two systems have the same number of satellites.
(3)   In the signal frequency band, the frequency difference between the two systems is not more than 30 MHz. Therefore, same antenna and bandwidth preamplifier can be shared to receive two signals.

(4)   In terms of positioning accuracy, GLONASS and GPS both provide two accuracy levels. The high accuracy is for military and special users, and the low accuracy is for the public users.

(5)   In the scope of application, both can be used for land, sea, air and space carrier navigation, positioning and timing.

Although GLONASS and GPS have many similarities, they are quite different in the key coordinate system, time system and modulation mode, mainly reflected in the following aspects.

(1)   Satellite orbits are different. The constellation of GPS and GLONASS has 6 orbital planes and 3 orbital planes, respectively. Their orbital altitudes are also different.

(2)   The time system is different. Although the two time systems belong to the atomic clock system, the GPS time system adopts the coordinated universal time UTC (USNO) of Washington, which is a continuous time system without leap seconds. The GLONASS time system adopts the coordinated universal time UTC (SU) of the Soviet Union, which is a discontinuous time system with synchronous leap seconds. GPS time = UTC + leap second, GLONASS time = UTC + 3.00 h. Therefore, the difference between GLONASS and UTC (SU) is only 3 h plus error less than 1 μs, and there is no leap seconds difference.

(3)   The reference coordinate system is different. GPS adopts WGS-84 coordinate system, while GLONASS adopts PZ-90 coordinate system.

(4)   The content of the navigation message is different. GPS broadcast the navigation ephemeris in the form of Kepler orbit elements, which is updated every two hours. GPS calculates the instantaneous position of GPS satellite in WGS-84 coordinate system according to Kepler orbit equation and considering satellite perturbation. GLONASS directly gives the position and velocity of the satellite in the reference epoch, as well as the perturbed acceleration of the sun and moon to the satellite. A set of ephemeris parameters are updated every 30 min to calculate the instantaneous position of GLONASS satellite in the PZ-90 coordinate system.

(5)   Satellite recognition is different. GPS uses code division multiple access (CDMA) mode, and each satellite uses the same carrier frequency to transmit signals. GLONASS uses frequency division multiple access (FDMA) mode, and each satellite uses different frequency to transmit signals. Before 2005, the frequency band of GLONASS satellite has exceeded the requirements of ITU. Russia started to implement the frequency change plan of GLONASS at the request of ITU, and completed the plan of frequency transfer in 2005. Like GPS transmitting signal using dual band, GLONASS signal is also transmitted by L1 and L2 band. The L2 band signal also uses special code modulation to ensure the use of military and special users.

The differences between GLONASS and GPS are shown in Table 1.6.

Table 1.6 Differences between GLONASS and GPS

Parameters	GPS	GLONASS
Number of constellation satellites	24	24
Number of track surfaces	6	3
Track height	20183 km	19130 km
Track radius	26560 km	25510 km
Operation cycle	11 h 58 min 00 s	11 h 15 min 40 s
Track inclination	55°	64.8°
Carrier frequency	L1: 1575.42 MHz	L1: 1598.0625–1606.5 MHz
	L2: 1227.60 MHz	L2: 1242.9376–1249.6 MHz
Transmission mode	Code division multiple access (CDMA)	Frequency division multiple access (FDMA)
Modulation code	C/A code and P code	C/A code and P code
Satellite ephemeris data format	Kepler orbit parameters	Geocentric rectangular coordinate system parameters
System time reference frame	UTC(USNO)	UTC(SU)
Coordinate system	WGS-84	PZ-90
Code rate	C/A code: 1.023 mb/s	C/A code: 0.511 mb/s
	P Code: 10.23 mb/s	P Code: 5.11 mb/s
SA	Released	None
Navigation message format	Each message has a 30 s main frame, five 6 s sub frames for each main frame and 30 bits for each sub frame	Each message has a 150 s super frame, including five 30 s frames, fifteen 2 s strings per frame and 100 bits per string
Message delivery rate	500 baud	50 baud
Super frame time length	2.5 min	12.5 min

1.1.4   Galileo System

Galileo system is a global satellite navigation system that will be completed soon. The system is jointly built by the European Space Agency (ESA) and the European navigation satellite system administration, whose headquarters are located in Prague, Czech Republic. The purpose of the system is to provide an independent high-precision positioning system for EU countries. The system provides free basic services, but the high-precision positioning service is only provided to specific users. Its function is to provide positioning service within 1 m in horizontal and vertical direction, and provide better positioning service than other systems in high latitude area.

The first experimental satellite of Galileo GIOVE-A was launched on December 28, 2005. The first official satellite was launched on August 21, 2011. Until December

Table 1.7 Parameters of galileo system

Satellite parameters	Parameter value
Number of satellites per orbit	10 (9 working, 1 spare)
Number of satellite distributed orbital planes	3
Track inclination angle	56°
Track height	23,616 km
Operation cycle	14 h 4 min
Satellite lifetime	20 year
Satellite mass	625 kg
Electricity supply	1.5 kW
Radio frequency	1202.025, 1278.750, 1561.098, 1589.742 MHz

2016, Galileo had 18 satellites in orbit, which were put into operation on December 15, 2016, and provided free basic services. On December 13, 2017, the 19th to 22nd Galileo satellites were successfully launched.

1.   Composition of Galileo system

Galileo system is divided into three parts: space segment, ground segment and user segment. Like other GNSS, Galileo adopts the principle of time-distance measurement for navigation and positioning.

1)   Space segment

The satellite constellation of Galileo system is composed of 30 MEO satellites distributed in three orbits. The specific parameters are shown in Table 1.7.

The number of satellites in Galileo satellite navigation system is similar to the arrangement of satellites and the constellations of GPS and GLONASS. The working life of Galileo system is 20 years. And that of MEO constellation is designed to be 15 years. There are two types of satellite clock in Galileo system: passive hydrogen maser clock and rubidium clock. Under normal working conditions, the passive hydrogen maser clock will be used as the main oscillator. The rubidium clock will operate as a backup, and monitor the operation of the passive hydrogen maser clock.

2)   Ground segment

The ground segment is composed of integrity monitoring system, satellite track measurement and control system, time synchronization system and system management center. The system has two ground control stations, Oberpfafenhofen near Munich in Germany and Fucino in Italy. There are 29 Galileo sensor stations distributed around the world. Moreover, there are 5 S-band uplink stations and 10 C-band uplink stations, which are responsible for the data exchange between the control center and satellites. The control center and the sensor station are connected by redundant communication network.

The ground control station is the core of the system. Its main functions are: controlling the Galileo satellite constellation, ensuring the synchronization of the atomic clocks on the satellites with the system time, providing signal integrity information, monitoring the satellites status and services provided by the satellites, processing the internal and external information of the system, etc.

3)   User segment

Galileo system can provide users with a variety of services, including the following.

(1)   Open service. The public service of Galileo is similar to GPS, providing free positioning, navigation and timing signals. This service is aimed at popular applications, such as all kinds of smart phone terminals and vehicle navigation terminals.

(2)   Business services. Compared with public service, commercial service has additional functions: distributing encrypted additional data in open server; very accurate local differential application, using open signal to cover PRS signal E6; supporting Galileo system positioning application and good pilot signal of wireless communication network.

(3)   Life safety services. The effectiveness of life safety service exceeded 99.9%. The combination of Galileo system with current GPS, or with the new generation of GPS III and EGNOS, will be able to meet higher requirements. Life safety services will also be applied to ship arrival, locomotive control, vehicle control, robot technology, etc.

(4)   Public franchise services. Public franchised services will provide a wider range of continuous services to the European Community at a dedicated frequency. It mainly includes: (1) safeguarding European national security, such as emergency services, government work and official duties; (2) emergency rescue, transportation and telecommunications applications, (3) other economic and industrial activities of strategic significance to Europe.

2.   Signal characteristics of Galileo system

The 10 signals provided by Galileo system are distributed in three frequency bands, namely E5A and E5B (1164–1215 MHz), E6 (1215–1300 MHz) and L1 (1559–1592 MHz).

(1)   Two signals are modulated on E5A, including low rate navigation information and auxiliary navigation information. Two signals are modulated on E5B, including navigation information, integrity information, SAR data and auxiliary navigation information.

(2)   E6 modulates three signals, including encrypted navigation signal, commercial signal and auxiliary navigation information.

(3)   Three signals are modulated on L1, including encrypted navigation signal, navigation information, integrity information, SAR data and auxiliary navigation information.

3.   Characteristics of Galileo system

(1)   The power of satellite transmitting signal is high. Galileo system has higher signal power than GPS. It can complete positioning in some areas where GPS cannot. If users in a certain area need additional services, Galileo system can also be provided by virtual satellite. When the signal received by the user does not meet the positioning requirements (four different satellite signals), it can be supplemented by the satellite signal transmitted by the virtual satellite.

(2)   TCAR (three carrier phase ambiguity resolution) technology. The carrier phase measurement and positioning principle of Galileo system is the same as that of GPS, but Galileo satellite navigation system has at least three carrier frequencies. The European Space Agency (ESA) proposes a TCRA scheme using three carriers, which can solve the problem of integer ambiguity.

(3)   System communication. Galileo system is equipped with communication function at the initial stage of operation. It plans to realize its communication function through the existing communication network on the ground, mainly considering the usage of European Universal Mobile Telecommunications System (UMTS). In this regard, experts put forward a plan of Galileo and S-UMTS cooperative system (GAUSS), in which the receiver can receive and process communication signal and navigation signal at the same time, and has communication and navigation functions.

(4)   SAR services. Galileo system also provides a kind of search and rescue service (SAR). The service is completed by user receivers and satellites. The user sends rescue signal to the Galileo satellite. The signal is sent to COSPAS/SARSAT GEO satellite by Galileo satellite, and then transmitted to the ground rescue system. The ground station rescue system receives the rescue signal and feeds back the information to the user after confirmation rescue operations.

1.2 Several Common Signals

1.2.1 BDS Signal

1.   Signal structure of BDS-1 satellite

1)   Offest quadrature phase shift keying (OQPSK) modulation

The signal of BDS-1 satellite is modulated by OQPSK. OQPSK is a constant envelope digital modulation technology developed after QPSK, which is an improved QPSK modulation method. When the polarity of only one channel of data in I and Q channels changes, the phase of QPSK signal changes by 90°. When the two channels change at the same time (such as from 00 to 11), the phase of QPSK signal changes by 180°.

OQPSK can solve the phase mutation problem of QPSK. OQPSK staggers the data streams of the same direction branch (I) and the orthogonal branch (Q) by half a code period in time. The data stream of each branch goes through differential coding, then carries out B/SK modulation respectively. Finally, each stream outputs the OQPSK signal through vector synthesis by the synthesizer.

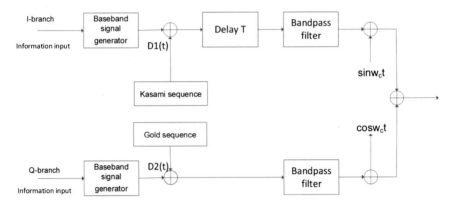

Fig. 1.1 Signal principle and structure of BDS-1 satellite

2)   Signal structure of BDS-1 satellite

The signal structure of BDS-1 satellite is shown in Fig. 1.1. Firstly, the information of I-branch and Q-branch is encoded by baseband signal generator respectively, which includes adding CRC check bit and data frame header to generate data stream with 8 kb/s. Then, convolutional coding method with encoding efficiency of 1 / 2 (2,1,7) and encoding length of 7 is adopted to generate non return to zero bipolar signal with 16 kb/s. It is multiplied by Kasami sequence and Gold sequence with encoding rate of 4.08 Mb/s to generate spread spectrum signal. The spread spectrum signal of I-branch is sinusoid modulated by B/SK after half a chip time delay; the spread spectrum signal of Q-branch is cosine modulated by B/SK directly, and finally added together to the channel.

2.   BDS-2 signal

1)   Signal frequency of BDS-2 satellite

At the ICG working group meeting on the compatibility of future navigation systems in Vienna in July 2009, China applied to ITU for multiple frequency bands, namely B1, B2 and B3, to transmit B1-C, B1, B2, B3 and B3-A navigation signals. Among them, B1 band is 1559–1563 MHz and 1587–1591 MHz, which overlap with E2 and E1 band of Galileo E2-L1-E1 band respectively; B2 band is 1164–1215 MHz, which partially overlap with E6 band of Galileo satellite.

2)   Signal characteristics of BDS-2 satellite

China National Administration of GNSS and Applications (CNAGA) announced the mode and frequency band used by BDS-2 satellite at the working group meeting in Vienna. BDS-2 satellite will widely use BOS modulation and its derivatives. Because the center frequency of B2a is 1176.45 MHz, BDS-2 receiver can be compatible with Galileo E5a signal or GPS L5 signal.

BDS-2 provides two types of services: Authorized Service (AS) and Open Service (OS). The AS service provides high-precision navigation and positioning at a higher security level, and contains the integrity information of the system. The service is designed for paid and military users. The OS service is free for global users, and its indicators are: positioning accuracy of 10 m, speed measurement accuracy of 0.2 m/s, and time service accuracy of 50 ns.

1.2.2 GPS Signal

GPS signal is a modulated wave sent by GPS satellite to users for navigation and positioning. The modulated wave is a combination of ranging code and satellite navigation message. GPS signal includes three signal components: carrier (L1 and L2), data code (navigation message) and ranging code (C/A code and P(Y) code). The composition of GPS signal is shown in Fig. 1.2.

The reference frequency of GPS satellite is directly generated by the atomic clock on the satellite, which is 10.23 MHz. All components of the satellite signal are multiplication or division of the reference frequency.

1.   Carrier

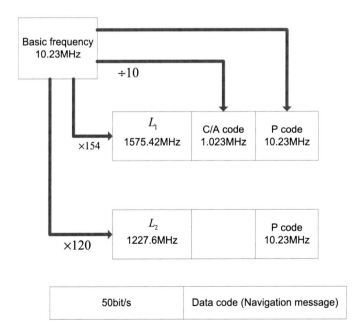

Fig. 1.2 Composition of GPS signal

Fig. 1.3 Carrier wavelength

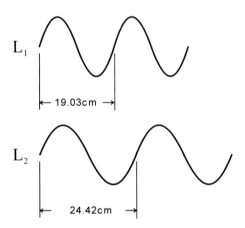

The main function of carrier is to carry other modulated signals, measure Doppler shift and ranging. L1 and L2 carriers are widely used. As shown in Fig. 1.3, the frequency of L1 is 1575.43 MHz and the wavelength is 19.03 cm; the frequency of L2 is 1227.60 MHz and the wavelength is 24.42 cm. Nowadays, L5 signal is added, with frequency of 1176.45 MHz and wavelength of 25.48 cm.

The carrier frequency used in GPS is helpful to weaken the influence of ionosphere refraction and to measure Doppler frequency shift. Selecting two frequencies can eliminate the ionosphere refraction delay of the signal (the frequency of the signal affects the ionosphere refraction delay).

2.   Ranging code

GPS satellites mainly generate two kinds of ranging code, namely P(Y) code and C/A code, which belong to pseudo-random noise code. Because of the complexity of these two kinds of ranging codes, we only gives a brief description of their generation, characteristics and functions.

C/A code is used for addressing, rough ranging and searching satellite signals. It belongs to a civil code and has a certain anti-jamming ability. The C/A code is generated by the combination of two 10 level feedback shift registers. The generation principle of C/A code is shown in Fig. 1.4.

Two 10 stage feedback shift registers in C/A code generator are all set to "1" state under the action of "1" pulse at zero o'clock every Saturday. Under the drive of 1.023 MHz clock pulse, the two shift registers generate m-sequences with the length of $N = 1023$ and period of $Nt_0 = 1$ ms, respectively. Their characteristic polynomials are shown as follows.

$$\begin{cases} G_1 = 1 + x^3 + x^{10} \\ G_2 = 1 + x^2 + x^3 + x^6 + x^8 + x^9 + x^{10} \end{cases} \tag{1.1}$$

In order to make different satellites have different C /A codes, the output of the two shift registers apply a very special combination. The output sequence of $G_1(t)$

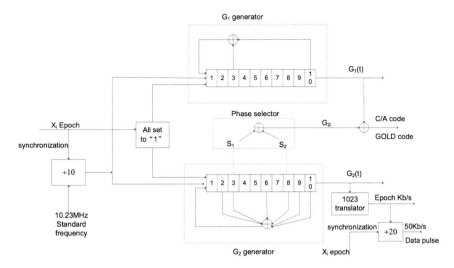

Fig. 1.4 C/A code generation principle

is provided directly. The states of two storage units are selected by $G_2(t)$ to modulus 2 add and then output, so that an m sequence G_{2i} equivalent to $G_2(t)$ translation can be obtained. Then it is modulus 2 added with $G_1(t)$ to generate C/A codes (also known as Gold codes) with different structures. Because $T = Nt_0 = 1$ms has 1023 bits, there may be 1023 equivalent translation sequences. After modulus 2 adding 1023 equivalent translation sequences to $G_1(t)$, 1023 kinds of m sequences can be generated, that is, 1023 kinds of C/A codes with different structures. This can completely cover the address demand of 24 satellites.

The code length, code rate and period of this group of C/A codes are the same, that is, the code length is $N = 2^{10} - 1 = 1023$bit, the code width is $t_0 = 1/f = 0.97752\mu s$ (the distance is about 293.1 m), the code period is $T = Nt_0 = 1$ms, and the code rate is 1.023 MHz.

Because the length of C/A code is very short, only 1023 bit, it is easy to capture. In order to capture the C/A code and determine the propagation delay of satellite signal, we usually need to search the C/A code one by one. If the C/A code is searched at the rate of 50 codes per second, the search time is only 20.5 s for the C/A code with 1023 codes. After using the C/A code to capture the satellite, we can obtain the navigation message. Through the information of the navigation message, we can easily capture the P(Y) code of GPS. Therefore, C/A code is also called capture code.

The code width of C/A code is wide. Assuming that the code error of two sequences is 1/100–1/10 of the code width, the ranging error of C/A code is 2.93–29.3 m. C/A code is also called coarse code because of its low precision.

P code is a kind of precise code, which is mainly used in military. When the AS is started, the P code is encrypted to form the Y code. The chip rate of P code is the same as that of Y code, which is usually abbreviated as P(Y). The two groups

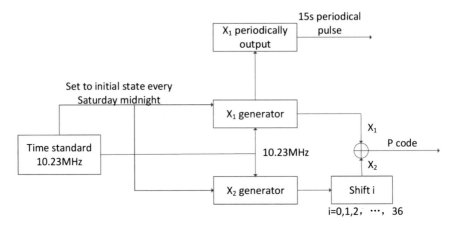

Fig. 1.5 The principle of generating P(Y) code

each have two 12 level feedback shift registers, which are combined to generate P(Y) code. The principle of generating P(Y) code is shown in Fig. 1.5.

The total number of m sequence generated by 12 stage feedback shift register is $2^{12} - 1 = 4095$. By truncation method, two 12 level m sequences are truncated into a truncated code with prime number of code in one cycle. For example, the number of X_{1a} and X_{1b} code is 4092 and 4093, respectively. The sum of X_{1a} and X_{1b} is modulus 2 added, and the long period code with period of 4092 × 4093 is obtained. Then, the product code is truncated, and X_1 with a period of 1.5 s and a number of code N_1 equal to 15.345 × 106 is cut out. The principle of generating X_1 code is shown in Fig. 1.6.

Through the same steps, in the other group, two 12 stage feedback shift registers generate X_2 codes, only the code period is slightly longer than that of X_1, $N_2 = 15.345 \times 10^6 + 37$(bit).

The code number of the product of N_1 and N_2 code is $N = N_1 \cdot N_2 = 23546959.765 \times 10^3$bit and the corresponding period is $T = N/10.23 \times 10^6 \times 86400 = 266.4$(d) ≈ 38(week).

The product code is $X_1(t) \cdot X_2 t + i \times t_0$, t_0 is the code width, i can be taken as 37 kinds of numerical values: 0, 1, …, 36, to generated 37 kinds of product codes. By intercepting a period of one week in the product code, 37 kinds of P(Y) codes with the same period and different structures (all are one week) can be generated. For the 24 GPS satellites, each satellite can use one of 37 kinds of P(Y) codes. Therefore, the P codes used by each satellite are different from each other, which is typically CDMA. Among the 37 P codes, 5 are used by ground stations and 32 are used by GPS satellites. Every Saturday at zero o'clock, the initial state of X_1 and X_2 will be set to "1". After one week, the state will return to the initial. Because of the length of P code sequence, it takes 14×10^5 days to search with speed of 50 codes /s, which is not practical. Therefore, the C/A code is usually captured at first, and then the P(Y) code is captured according to the information given in the navigation message.

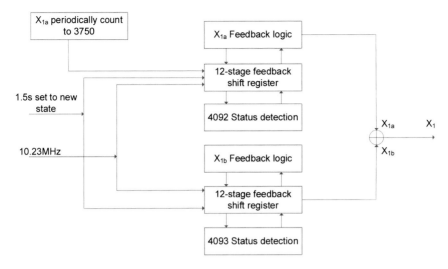

Fig. 1.6 The principle of generating codes X_1

Since the code width of P(Y) code is only 1/10 of that of C/A code, if the alignment accuracy of the code is still 1/100–1/10 of the code width, the ranging error is 0.2936–2.936 m, which is 10 times higher than that of C/A code. Therefore, P(Y) code can be used for more precise positioning, usually also known as precise code.

Due to its long period (7 days), various code types and high code rate (10.23 MHz), P(Y) code is a military code for precise ranging, anti-jamming and secrecy. According to the regulations of the US Department of Defense, P(Y) code is exclusively for military. At present, only licensed user receiver can receive P(Y) code, and the price is expensive. Therefore, the development and research of codeless receiver, Z technology and square technology to fully develop GPS information resources has become a work of great practical value.

3.   GPS navigation message

1)   Navigation message format

Navigation message refers to the data code containing navigation information. Navigation information includes satellite ephemeris, satellite almanac, satellite working status, clock correction parameters, time system, atmospheric refraction correction parameters, orbit perturbation correction parameters, telemetry code and exchange code of P(Y) code determined by C/A code, etc.. Navigation information is the data basis for users to apply GPS for navigation and positioning.

Navigation message is a binary code file, which is composed of data frame with certain format and broadcast out. The navigation message format is shown in Fig. 1.7. Each frame contains five subframes, which contain 1500 bits. Each subframe contains 10 words and each word is 30 bit. The broadcast speed of navigation message is 50 b/s, and the broadcast time of each subframe is 6 s.

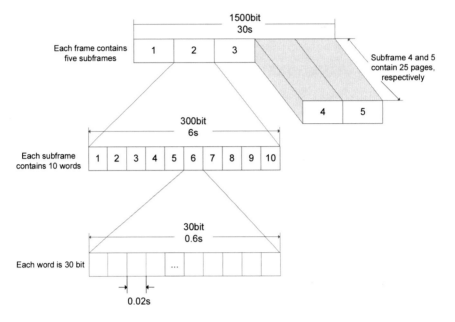

Fig. 1.7 Navigation message format

Subframe 1, 2 and 3 cycle once every 30 s. Subframe 4 and 5 have 25 forms, each containing 25 pages. Subframe 1, 2, 3 and each page of sub frame 4 and 5 constitute a frame. The whole navigation message has 25 frames and 37,500 bits, which takes 12.5 min to completely broadcast.

Subframe 1, 2 and 3 contain the satellite clock correction parameters and broadcast ephemeris of the satellite, and their contents are updated every hour. Subframe 4 and 5 are ephemeris of all GPS satellites, and their contents are updated only after the ground station uploads new navigation data.

2) Content of navigation message

The content of each subframe in each navigation message is shown in Fig. 1.8. Each subframe is composed of handover word (how), telemetry word (TLW) and data block. The third to tenth words of 1, 2 and 3 subframes constitute data block II, and the third to tenth words of 4 and 5 subframes constitute data block III.

(1) Telemetry word (TLW): telemetry word is the first word of each subframe, which is used as the preamble to capture navigation message and provides a starting point for synchronization of each subframe. TLW has 30 bits in total, the frame header (synchronization code) is 1–8 bits; telemetry message is 9–22 bits, including the status information, diagnosis information and other information when the ground monitoring system uploads data; reserved bit is 23 and 24 bit; parity check bit is 25–30 bit.

(2)   Hand over word (how): the hand over word is the second word of each subframe, with a total of 17 bits. It is the time count (positive count) from midnight every Saturday/Sunday. Users can capture P code quickly. The 18th bit indicates whether the rolling momentum moment of the satellite is unloaded after the information is uploaded. The 19th bit indicates the synchronization of the satellite, indicating whether the time of the data frame is consistent with the time of the code clock. The 20th to 22th bits are the marks of the subframe identification.

(3)   Data block I: data block I mainly contains health status data and satellite clock, including:

① Satellite time counter (WN): the number of weeks from UTC zero time on January 5, 1980 is called GPS week, which is located in the first to tenth bit of the third word.

② Modulation code identification: 11–12 bit of the third word, "10" is C/A code modulation, "01" is P code modulation.

③ User ranging accuracy (URA): 13–16 bit of the third word, "10" is C/A code modulation, "01" is P code modulation.

④ The 17th bit of the third word indicates whether the navigation data is normal, and the 18th to 22th bits indicate the correctness of signal coding.

⑤ Ionospheric delay correction parameter (T_{GD}): it corrects the ionospheric delay difference of L1 and L2 carriers, occupying the 17th to 24th bit of the 7th word, and provides rough ionospheric refraction correction for users of single frequency receivers (this correction is not required for dual frequency receivers).

⑥ Age of data, clock (AODC): the extrapolation time interval of clock correction, which is the difference between the reference time of satellite clock correction parameter t_{oc} and the last measurement time of calculating the correction parameter t_L, i.e. AODC $= t_{oc} - t_L$.

⑦ Satellite clock correction parameters: the clock on each satellite is corrected according to GPS time. Although the GPS satellite clock uses high precision cesium atom clock and rubidium atom clock, there is still deviation. In addition, due to the relativistic effect, the satellite clock moves faster than the ground clock, with a difference of 448 ps per second (3.87×10^{-5}s per day). We reduce the nominal frequency of the satellite from 10.23 MHz to the actual frequency of 10.2299999545 MHz to eliminate this effect. However, the time offset caused by the relativistic effect is not constant. Besides, the quality of each clock is also different. Therefore, there is an error between the time indicated by the satellite clock and the ideal GPS time, which is called the satellite clock error.

$$\Delta t = a_0 + a_1(t - t_{oc}) + a_2(t - t_{oc}) \tag{1.2}$$

where, a_0 is the deviation (zero deviation) of the satellite clock from the GPS time at the reference time t_{oc} of the satellite clock; a_1 is the frequency deviation (clock speed) of the satellite clock relative to the actual frequency at the reference time t_{oc} of the satellite clock; a_2 is the drift coefficient (clock drift) of the satellite clock frequency. t_{oc} occupies 9–24 bit of the 8th word, a_0 occupies 1–22 bit of the 10th word, a_1 occupies 9–25 bit of the 9th word and a_2 occupies 1–8 bit of the 9th word.

(4)   Data block II: data block II, also known as satellite ephemeris, is the core part of navigation message. The data block II contains the calculating information of the position of the satellite. The GPS receiver performs real-time navigation and positioning calculation according to the parameters of the satellite ephemeris. The satellite sent this block every 30 s and updated every hour.

(5)   Data block III: data block III contains the almanac data of all GPS satellites, which is the general form of each satellite ephemeris. The main contents are as follows: ① Page 1–24 of the 5th subframe provides the almanac of 1–24 satellites; ② Page 25 of the 5th subframe provides the health status and GPS week number of 1–24 satellites; ③ Page 2–10 of the 4th subframe provides the almanac of satellite 25–32; ④ Page 25 of the 4th subframe provides the anti-spoofing characteristics of 32 satellites (AS off or on) and the health status of 25–32 satellites; ⑤ Page 18 of the 4th subframe provides the ionosphere delay correction model parameters α_0, α_1, α_2, α_3, β_0, β_1, β_2, β_3 and the correlation parameter Δt_G between GPS time and UTC, which are calculated by the following equation,

$$\Delta t_G = A_0 + A_1(t - t_{01}) + \Delta t_{LS} \tag{1.3}$$

where, t_{01} is the reference time; Δt_{LS} is the time change caused by the second jump.

When the GPS receiver captures a certain satellite, it can quickly capture other satellite signals and select the most suitable satellite by using the almanac, code division address, clock correction and satellite status provided by data block III.

1.2.3 GLONASS Signal

1.   GLONASS signal structure

Like GPS satellite, GLONASS also transmits L1 and L2 carrier signals, and uses B/SK modulation on the carrier with navigation message for positioning and pseudo-random code for ranging. Different from GPS code division multiple access (CDMA) multiplexing technology, GLONASS uses frequency division multiple access (FDMA) mode. Each satellite transmits the same PRN code at different frequencies. The receiver can tune the receiving frequency to the desired satellite frequency according to the certain satellite signal.

Because of the complexity of the front-end components, FDMA is usually expensive and will increase the volume of the receiver, while CDMA signal processing can share the same front-end components. But the anti-jamming ability of FDMA is obviously stronger. Generally, the interference signal source can only interfere with one FDMA signal, and FDMA does not need to consider the interference effect (cross-correlation) between multiple signals. Therefore, GLONASS has more anti-jamming options than GPS, and has simpler code selection criteria.

The generation principle of GLONASS satellite signal is shown in Fig. 1.9. Each satellite transmits signal with two separate L1 and L2 carriers as the center. Similar to GPS, the PRN ranging code of GLONASS is also composed of military P code and civil C/A code. The difference is that GLONASS contains two kinds of navigation messages, corresponding to P code and C/A code respectively. P code $\oplus$ P code message, C/A code $\oplus$ C/A code message are modulated on L1 carrier, and P code $\oplus$ P code message is modulated on L2 carrier. After GLONASS modernization, GLONASS-M satellite adds navigation message function. In order to improve the navigation accuracy of civil satellite, C/A code is also modulated on L2 carrier.

2. GLONASS signal frequency

GLONASS satellite adopts FDMA mode. According to the initial design of the system, the frequencies of L1 and L2 carrier signals sent by each satellite are different from each other. Each GLONASS satellite determines the corresponding carrier frequency (MHz) according to the following equation:

$$f = (178.0 + K/16) \cdot Z \tag{1.4}$$

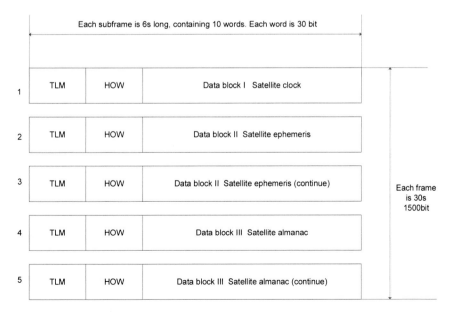

Fig. 1.8 Content of frame navigation message

where, K is the frequency of the signal transmitted by GLONASS satellite, taking a positive integer; Z is a multiple parameter; the L1 carrier is taken as 9, and the L2 carrier is taken as 7. Therefore, it can be further obtained that the carrier frequency of each GLONASS satellite is

$$
\begin{aligned}
f_{L1} &= 1602 + 0.5625 \cdot K \\
f_{L2} &= 1246 + 0.4375 \cdot K
\end{aligned}
\tag{1.5}
$$

It can be seen from Eq. (1.5) that the adjacent frequency interval on L1 band is 0.5625 MHz, and that on L2 band is 0.4375 MHz.

With the development of GLONASS satellite technology, its frequency plan has been changed. At the beginning of the system design, the channel K of GLONASS satellite is 0–24, which can identify 24 satellites. However, there is a certain cross interference between the frequency obtained and the frequency of radio astronomy research (1610.6–1613.8 MHz). In addition, the International Telecommunication Union (ITU) has allocated the frequency band 1610.0–1626.5 MHz to the near earth satellite mobile communication. Therefore, Russia plans to reduce the frequency and carrier bandwidth of GLONASS satellite. The frequency modification plan is divided into two steps: 1998–2005, channel number $K = -7$ to 12; after 2005, channel number $K = -7$ to 4.

After the frequency change, the final configuration will only use 12 channels (K $= -7$ to 4). But there are 24 satellites, it is planned to let the satellites on both sides of the earth share the same K value. Because anywhere on the earth, it is impossible to see two satellites with 180° difference in position on the same orbital plane at the same time. These two satellites can use the same frequency without interference. The frequency plan is the recommended value under normal conditions. Russia may also allocate other K values for some special situations such as military command or control.

3.   GLONASS code characteristics

GLONASS satellite is similar to GPS satellite in that it adopts pseudo-random code, which is convenient for ranging. Each satellite uses two PRN codes to modulate its L-band carrier. One is called P code for military usage, and the other is called C/A code for civilian usage, which can assist in the acquisition of P code. Because GLONASS satellite adopts FDMA mode, its specific pseudo-random code design and characteristics are different from GPS satellite.

The C/A code of GLONASS satellite uses the longest 9-stage feedback shift register to generate PRN code sequence. The code repetition period is 1 ms, the code length is 511 bit, and the code rate is 0.511 Mb/s.

The C/A code of GLONASS satellite uses this relatively short code at high clock rate. The main advantage is that it can be acquired quickly, and the high code rate is conducive to enhancing the resolution of long distance. The disadvantage is that the short code will produce some unwanted frequency components at the frequency of 1 kHz, resulting in cross-correlation with interference sources, thus, weakening the anti-jamming performance of spread spectrum. However, since the frequencies of GLONASS satellites are separated, the correlation between satellite signals can be significantly reduced.

The P code of GLONASS satellite uses the longest 25 stage feedback shift register to generate PRN code sequence. The code rate is 5.11 Mb/s, and the repetition period is 1 s (the actual repetition period is 6.57 s, but the chip sequence is truncated once every 1 s).

Compared with C/A code, P code only repeats once per second. Although it produce unwanted frequency components at 1 Hz intervals, its related problems are not as serious as C/A code. Similarly, FDMA technology actually eliminates the cross-correlation problem between satellite signals. Although P code has advantages in correlation characteristics and confidentiality, it has made sacrifices in acquisition efficiency. The P code has the possibility of 5.11×10^8 code phase shift. Therefore, the receiver should capture the C/A code first, and then assist to capture the P code according to the C/A code.

4. GLONASS navigation message

Different from GPS, the navigation message of GLONASS satellite consists of P code and C/A code. The data streams of the two kinds of navigation messages are 50 b/s, and are modulated in the form of modular 2 addition to P code and C/A code, respectively. Navigation message is mainly used to provide channel allocation information and satellite ephemeris, as well as satellite health status, epoch timing synchronization bit and other information. In addition, Russia also plans to provide data conducive to the integrated use of GLONASS and GPS, such as the difference between WGS-84 and PZ-90, and the system time difference between the two satellite navigation systems.

1) C/A code navigation message

The navigation message of GLONASS satellite is transmitted according to Hamming code, which is a binary code. A complete navigation message is generally a super-frame composed of five frames. Each frame contains 15 lines and each line is 100 bit. Figure 1.10 shows the C/A code navigation message format of GLONASS satellite. The broadcast repetition time of each frame is 30 s, and the broadcast time of the whole navigation message is 2.5 min.

The first three rows of each frame contain satellite real-time data, including satellite orbit parameters, detailed ephemeris of the tracked satellite and satellite clock correction parameters, etc. The other rows contain general ephemeris information of other satellites in GLONASS constellation, as well as non real-time data such as approximate time correction and health status of all satellites, among which each frame contains ephemeris of five satellites.

2) P code navigation message

As the P code is military code, Russia did not disclose the details of the P code message. Some independent research institutions have published the characteristics

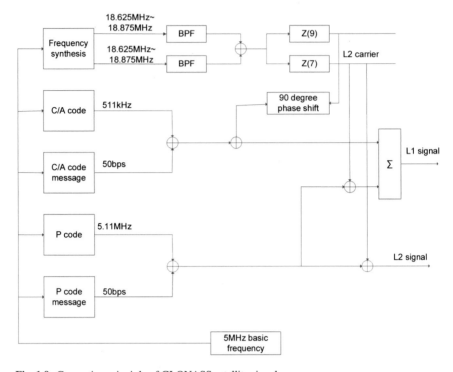

Fig. 1.9 Generation principle of GLONASS satellite signal

of some P codes by studying the received GLONASS satellite signals. These information cannot guarantee its continuity and other aspects. Russia may adjust the P code at any time without prior notice.

P code navigation message is a superframe composed of five frames, each frame contains five lines, each line is 100 bit. The broadcast time of each frame is 10 s, and the broadcast time of the whole navigation message is 12 min. The first three lines of each frame contain the detailed information of the tracked satellite, and the other lines contain the approximate ephemeris of other satellites in GLONASS constellation.

The biggest difference between P code message and C/A code message is that it takes 12 min and 10 s for the former to obtain all satellite approximate ephemeris and real-time ephemeris respectively, while the latter takes 2.5 min and 30 s respectively.

1.2.4 Galileo Signal

1.   Galileo frequency planning

Galileo system is mainly designed to meet the needs of different users. It defines five basic services independent of other satellite navigation systems: Open Service

(OS), Safety-of-Life service (SOL), Commercial Service (CS), Public Regulated Service (PRS) and Search and Rescue support service (SAR). It transmits different types of data in different frequency bands, therefore, Galileo is a multi-carrier satellite navigation system.

Galileo system will provide six kinds of right-handed circularly polarized (RHCP) navigation signals in E5 band (1164–1215 MHz), E6 band (1260–1300 MHz) and E2-L1-E1 band (1559–1300 MHz). Among them, E5 band can be divided into E5a and E5b bands. E2-L1-E1 band is an extension of L1 band of GPS satellite, which can also be expressed as L1 for convenience.

All the frequency bands of Galileo system are located in the radio navigation satellite service (RNSS) frequency band. At the same time, E5 and L1 frequency bands have been assigned to the aeronautical radio navigation service (ARNS). Therefore, the signals in this frequency band can be applied to the special aviation related services with high safety requirements.

The center frequency of L1 band (E2-L1-E1) is 1575.42 MHz, which is the same as that of GPS L1 band. The center frequencies of E5a and E5b bands are 1176.45 MHz and 1207.14 MHz, respectively. This is to maintain compatibility with GPS satellites.

2. Galileo signal design

Galileo signal is named by its frequency band. Each satellite will transmit six kinds of navigation information: L1F, L1P, E6C, E6P, E5a, E5b, and L6 signal specially used for SAR. Various signals are described as follows:

(1) L1F signal: located in L1 band. It is a publicly accessible signal, including a none data channel (called pilot channel) and a data channel. It modulates unencrypted ranging code and navigation message, which can be received by all users. In addition, it also contains encrypted business information and integrity information.

(2) L1P signal: located in L1 band. It is a restricted access signal, and its message and ranging code are encrypted by official encryption algorithm.

(3) E6C signal: located in E6 band. It is a signal for commercial access, including a pilot channel and a data channel. Its ranging code and message are encrypted by commercial encryption algorithm.

(4) E6P signal: located in E6 band. It is a restricted access signal. Its message and ranging code are encrypted by official encryption algorithm.

(5) E5a signal: located in E5 band. It is a publicly accessible signal, including a pilot channel and a data channel. It modulates unencrypted ranging code and navigation message, which can be received by all users. The basic data transmitted is used to support navigation and time service functions.

(6) E5b signal: located in E5 band. It is a publicly accessible signal, including a pilot channel and a data channel. It modulates navigation message and unencrypted ranging code, which can be received by all users. In addition, the data stream also contains encrypted commercial data and integrity information.

(7)   L6 signal: retrieve the distress information in the frequency band of 406–
      406.1 MHz, and transmit it to the special ground receiving station in the
      frequency band of 1544–1545 MHz (reserved for emergency service).

3.   Galileo spread spectrum code

Galileo signal not only adopts new modulation system, but also applies new technology in its spread spectrum code. The spread spectrum code (ranging code) used
in Galileo signal can be divided into main code and sub code. The former is used in
pilot channel and data channel at the same time, while the latter is only used in pilot
channel. The main code is a pseudo-random code which is usually used in spread
spectrum of satellite signal, and the sub code is an innovation of Galileo signal. It
modulates the signal again on the basis of the main code to form a layered structure
code. The main code generator is based on the traditional gold code. Its linear feedback shift register can reach up to 25 levels, and the predefined sequence length of
the sub code is up to 100 bits. At present, the final code parameters of Galileo signal
are still in the stage of experiment and optimization.

The spread spectrum code design of Galileo signal provides a good tradeoff
between anti-jamming protection and acquisition time. For the received satellite
signal, when the signal-to-noise ratio is high, only the main code needs to be correlation despreading to obtain the required correlation gain; when the signal-to-noise
ratio is low, the secondary code can be further correlation despreading to obtain the
further correlation gain.

4.   Galileo navigation message

Galileo navigation message adopts a fixed frame format, which makes the distribution
of given message data content (integrity, almanac, ephemeris, clock correction, ionosphere correction, etc.) on the subframe flexible. In order to improve the transmission
efficiency, the research of frame format for different signals is in progress.

The complete navigation message is transmitted in the form of super phase frame
on each data channel. A super phase frame contains several sub frames. The sub frame
is composed of data field, unique word (UW), cyclic redundancy check (CRC), tail
bit, etc.. The basic structure of navigation message is shown in Fig. 1.11.

The UW of the subframe can make the receiver complete the synchronization of
the data domain boundary, and the synchronization code at the transmitter adopts
the uncoded data symbols. The CRC check covers the data domain of the whole
subframe (except the tail bit and UW). After all the subframes are encoded by forward
error correction (FEC), all the subframes (excluding the synchronization code) are
protected by block interleaving.

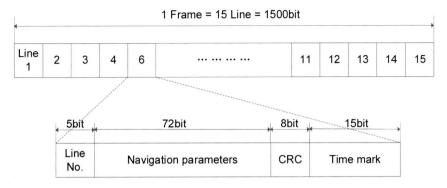

Fig. 1.10 C/A code navigation message format of GLONASS satellite

1.3 Principle of Satellite Navigation and Positioning

1.3.1 Pseudo Range Measurement

1. The concept of pseudo range

Pseudo range positioning is the most commonly used positioning method. According to the pseudo range of four or more GNSS satellites and the known satellites coordinates obtained by GNSS receiver at a certain time, the three-dimensional coordinates of the receiver antenna are obtained by using the spatial range intersection method. The measured pseudo range is the measured distance obtained by multiplying the propagation time of the signal transmitted from the satellite to the receiver by the propagation speed of the signal (the speed of light). Due to the influence of satellite clock, receiver clock, multipath effect and the delay of satellite signal passing through the ionosphere and troposphere, there is a certain difference between the actual measured distance and the geometric distance from the satellite to the receiver antenna. Therefore, the measured distance including the error is generally called pseudo range.

The distance between the receiver and the satellite can be calculated by measuring the propagation time of the ranging code signal transmitted from GNSS satellite to the receiver, as follows

$$\rho' = \Delta t \cdot c \tag{1.6}$$

where Δt is the propagation time and c is the speed of light.

The distance calculated by Eq. (1.6) is pseudo range ρ', and its relationship with geometric distance can be expressed as follows:

$$\rho' = \rho + \delta\rho_1 + \delta\rho_2 + c\delta t_i - c\delta t^j \tag{1.7}$$

where, $\delta\rho_1$ and $\delta\rho_2$ represent the ionospheric and tropospheric correction terms, respectively; δt_i represent the deviation of the receiver clock from the standard time; δt^j represent the deviation of the satellite clock from the standard time.

Although the single point positioning accuracy by pseudo range method is not high (the positioning error is about 10 m), it is still the basic navigation method of GNSS positioning system because the ranging code used in pseudo range method has the advantages of no ambiguity and fast positioning speed. At the same time, the pseudo range can be used as an auxiliary value to solve the integer period ambiguity in carrier phase measurement. Therefore, it is necessary to understand the basic principles and methods of pseudo range measurement and pseudo range positioning.

2.   Principle of pseudo range measurement

The key step in pseudo range positioning is pseudo range measurement. Its basic process is as follows.

GNSS satellite sends out the ranging code of a certain structure according to its own clock, and the ranging code arrives at the receiver after propagation time τ. Under the control of its own clock, the receiver generates a group of duplication codes with the same structure as the ranging codes sent by the satellite, and correlates the two groups of ranging codes by delayer τ'. If the self-correlation coefficient $R(\tau') \neq 1$, the delay time τ' will continue to be modulated until the self-correlation coefficient $R(\tau') = 1$. At this time, the received GNSS ranging codes and the duplication codes generated by the receiver are completed in full alignment, the delay time τ' is the time that GNSS satellite signal propagates from satellite to receiver. The distance from satellite to receiver is the product of τ' and c.

The principle of pseudo range measurement is shown in Fig. 1.12. The measurement of self-correlation coefficient $R(\tau')$ is completed by the integrator and correlator in the receiver's phase locked loop (PLL).The ranging code $a(t)$ controlled by the satellite clock is sent out from the satellite antenna at the time of GNSS time t, passes through the ionosphere and troposphere, and arrives at the GNSS receiver through time delay τ. The signal received by the receiver is $a(t - \tau)$.The local code generator controlled by the receiver clock will generate a local code $a(t + \Delta t)$ which is the same as the ranging code sent by the satellite. Δt is the clock difference between the receiver clock and the satellite clock. The code shift circuit delays the local code with time τ'. The correlator performs correlation operation on the received satellite signal. After passing through the integrator, the self-correlation coefficient $R(\tau')$ can be obtained.

$$R(\tau') = \frac{1}{T} \int a(t - \tau)a(t + \Delta t - \tau')dt \qquad (1.8)$$

where, T is the period of the ranging code. Adjusting the time delay τ', we can obtain the maximum correlation output value, to obtain the pseudo range ρ'. Figure 1.12 is the schematic diagram of pseudo range measurement.

The ranging codes generated and transmitted by each GNSS satellite are arranged according to certain rules, and the codes and time are one-to-one corresponding in one

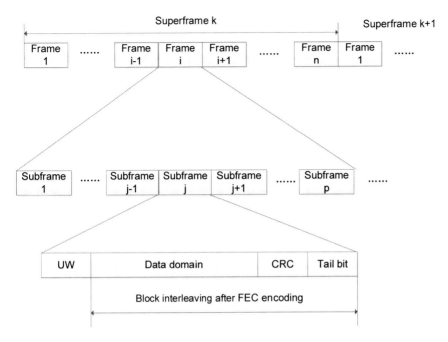

Fig. 1.11 Basic composition of Galileo navigation message

period. The GNSS receiver recognizes the shape characteristics of each code, and then calculates the signal propagation delay τ' to obtain the pseudo range. Therefore, the code correlation technology can be used to determine the pseudo range in the pseudo range measurement. But in fact, each code generated in the receiver has a certain random error. The satellite signal is affected by the error after long-distance transmission between the satellite and the receiver, which will also produce deformation. Therefore, it will produce large error to calculate the signal time delay τ' according to the shape characteristics of the ranging code. Using the code correlation technology to determine the propagation time τ' of the signal when the self-correlation coefficient $R(\tau')$ between the replication code and the ranging code reaches to the maximum, it effectively eliminates the influence of various random errors. In essence, it uses multiple ranging code features to determine the delay. Because the replication code of the receiver and the ranging code of the satellite signal inevitably have errors in the process of generation, and the ranging code of the satellite signal will be deformed due to various external interference in the process of transmission, the self-correlation coefficient of the two cannot reach to "1". Therefore, the local replication code and the received satellite ranging code can only be confirmed when their self-correlation coefficient is the maximum. The pseudo range is determined by basic alignment. In this way, the influence of various random errors can be eliminated to the greatest extent to improve the accuracy.

1.3.2 Pseudo Range Observation Equation and Positioning Calculation

1. Pseudo range observation equation

In GNSS positioning, the observation equation is mainly used to describe the functional relationship between observations and location parameters. The equation with distance delay of ranging code (C/A code or P code) as the observation is called pseudo range measurement observation equation, also known as pseudo range observation equation.

Before establishing the pseudo range measurement equation, we first make some symbolic provisions.

(1) t^j (GNSS): the ideal GNSS time when a satellite S^j transmits a signal;
(2) t_i (GNSS): the ideal GNSS time when the receiver T_i receives the satellite signal;
(3) t^j: the time of the satellite clock when the satellite S^j transmits the signal;
(4) t_i: the time of the receiver clock when the receiver T_i receives the satellite signal;
(5) Δt_i^j: the propagation time of the satellite signal arriving at the observation station;
(6) δt^j: clock difference of satellite clock relative to ideal GNSS time;
(7) δt_i: clock difference of receiver clock relative to ideal GNSS time.

There are

$$t^j = t^j(\text{GNSS}) + \delta t^j$$
$$t_i = t_i(\text{GNSS}) + \delta t_i \tag{1.9}$$

The time of signal transmission from satellite to receiver is

$$\Delta t_i^j = t_i - t^j = t_i(\text{GNSS}) - t^j(\text{GNSS}) + \delta t_i - \delta t^j \tag{1.10}$$

Assuming that the geometric distance between the satellite and the observation station is ρ_i^j, the corresponding pseudo range $\tilde{\rho}_i^j$ can be obtained without considering the atmospheric influence

$$\tilde{\rho}_i^j = \Delta t_i^j \cdot c = c \cdot \Delta \tau_i^j + c \cdot \delta t_i^j = \rho_i^j + c \cdot \delta t_i^j \tag{1.11}$$

where, $t_i(\text{GNSS}) - t^j(\text{GNSS}) = \Delta \tau_i^j$, $\delta t_i - \delta t^j = \delta t_i^j$. When the satellite clock and the receiver clock are strictly synchronized, $\delta t_i - \delta t^j = \delta t_i^j = 0$. The pseudo range determined by Eq. (1.11) is the geometric distance between the station and the satellite.

Usually, the clock error of GNSS satellite can be obtained from the navigation message transmitted by the satellite. After clock error correction, the time synchronization error between satellites can be kept within 20 ns. If the influence of satellite clock error is ignored and the refraction of ionosphere and troposphere is taken into account, the common form of pseudo range observation equation is as follows

$$\tilde{\rho}_i^j(t) = \rho_i^j(t) + c\delta t_i - c\delta t^j + \delta I_i^j(t) + \delta T_i^j(t) \tag{1.12}$$

where, $I(t)$ and $T(t)$ refer to ionospheric refraction correction and tropospheric refraction correction, respectively.

2. Positioning calculation

In Eq. (1.12), the tropospheric and ionospheric refraction correction terms can be calculated according to a certain model, and the satellite clock error can be obtained from the navigation message. Assuming that the refraction correction terms of troposphere and ionosphere have been accurately obtained, and the correction of satellite clock and receiver clock are known. Then once the pseudo range is determined, it is essentially equal to the geometric distance between stations. The relationship between geometric distance ρ_i, satellite coordinates (X_i, Y_i, Z_i) and receiver coordinates (X, Y, Z) is as follows:

$$\rho_i = \sqrt{(X_i - X)^2 + (Y_i - Y)^2 + (Z_i - Z)^2} \tag{1.13}$$

Since satellite coordinates can be obtained from satellite navigation messages, there are three unknowns in Eq. (1.13). If the user measures the pseudo range of three satellites at the same time, the position (X, Y, Z) of the receiver can be obtained.

In the above assumption, the clock correction at any observation moment is accurately known, which can only be realized by atomic clocks with excellent stability. It is possible to equip atomic clocks on a limited number of satellites. However, it is unrealistic to install atomic clocks on every receiver, which not only increases the volume and weight of the receiver, but also greatly increases the cost.

In order to solve the problem mentioned above, we take the clock correction of the receiver at the observation time as an unknown value. At any observation moment, the user needs to observe at least four satellites at the same time to calculate the four unknowns.

The geometric distance between the observation station and the satellite is nonlinear

$$\rho_i^j(t) = \sqrt{(X^j(t) - X_i)^2 + (Y^j(t) - Y_i)^2 + (Z^j(t) - Z_i)^2} \tag{1.14}$$

The pseudo range is obtained by substituting Eq. (1.14) into Eq. (1.12)

$$\tilde{\rho}_i^j(t) = \sqrt{(X^j(t) - X_i)^2 + (Y^j(t) - Y_i)^2 + (Z^j(t) - Z_i)^2} \\ + c\delta t_i - c\delta t^j + \delta I_i^j(t) + \delta T_i^j(t) \tag{1.15}$$

The pseudo range positioning method is shown in Fig. 1.13. The user observes satellites with labels of # 1, # 2, # 3, # 4 at the same time, and assumes that the

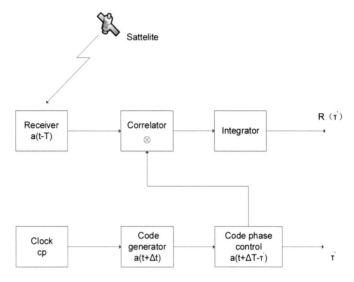

Fig. 1.12 Principle of pseudo range measurement

position coordinates of each satellite, (X_i, Y_i, Z_i), i $= 1, 2, 3, 4$, are known. It is assumed that the user's real position and its estimated position coordinates are (X, Y, Z) and (X_{es}, Y_{es}, Z_{es}), respectively.

$$\begin{cases} X = X_{es} + \Delta x \\ Y = Y_{es} + \Delta y \\ Z = Z_{es} + \Delta z \end{cases} \tag{1.16}$$

The distances of each satellite to the receiver estimated position are ρ_{es_1}, ρ_{es_2}, ρ_{es_3}, ρ_{es_4}, respectively.

$$\rho_{es\_1} = \sqrt{(X_1 - X_{es})^2 + (Y_1 - Y_{es})^2 + (Z_1 - Z_{es})^2} \tag{1.17}$$

$$\rho_{es\_2} = \sqrt{(X_2 - X_{es})^2 + (Y_2 - Y_{es})^2 + (Z_2 - Z_{es})^2} \tag{1.18}$$

$$\rho_{es\_3} = \sqrt{(X_3 - X_{es})^2 + (Y_3 - Y_{es})^2 + (Z_3 - Z_{es})^2} \tag{1.19}$$

$$\rho_{es\_4} = \sqrt{(X_4 - X_{es})^2 + (Y_4 - Y_{es})^2 + (Z_4 - Z_{es})^2} \tag{1.20}$$

By expanding Eqs. (1.17)–(1.20) with Taylor formula and substituting into Eq. (1.15), we can get

$$\tilde{\rho}_i^1 = \rho_{es\_1} + \frac{\partial \rho_{es\_1}}{\partial X_1} \cdot \Delta x + \frac{\partial \rho_{es\_1}}{\partial Y_1} \cdot \Delta y + \frac{\partial \rho_{es\_1}}{\partial Z_1} \cdot \Delta z + c\delta t_i^1 + \delta I_i^1(t) + \delta T_i^1(t)$$

$$(1.21)$$

$$\tilde{\rho}_i^2 = \rho_{es\_2} + \frac{\partial \rho_{es\_2}}{\partial X_2} \cdot \Delta x + \frac{\partial \rho_{es\_2}}{\partial Y_2} \cdot \Delta y + \frac{\partial \rho_{es\_2}}{\partial Z_2} \cdot \Delta z + c\delta t_i^2 + \delta I_i^2(t) + \delta T_i^2(t)$$

$$(1.22)$$

$$\tilde{\rho}_i^3 = \rho_{es\_3} + \frac{\partial \rho_{es\_3}}{\partial X_3} \cdot \Delta x + \frac{\partial \rho_{es\_3}}{\partial Y_3} \cdot \Delta y + \frac{\partial \rho_{es\_3}}{\partial Z_3} \cdot \Delta z + c\delta t_i^3 + \delta I_i^3(t) + \delta T_i^3(t)$$

$$(1.23)$$

$$\tilde{\rho}_i^4 = \rho_{es\_4} + \frac{\partial \rho_{es\_4}}{\partial X_4} \cdot \Delta x + \frac{\partial \rho_{es\_4}}{\partial Y_4} \cdot \Delta y + \frac{\partial \rho_{es\_4}}{\partial Z_4} \cdot \Delta z + c\delta t_i^4 + \delta I_i^4(t) + \delta T_i^4(t)$$

$$(1.24)$$

Substituting Eqs. (1.17)–(1.20) into Eqs. (1.21)–(1.24), we get

$$\tilde{\rho}_i^1 = \rho_{es\_1} + \frac{X_{es} - X_1}{\rho_{es\_1}} \cdot \Delta x + \frac{Y_{es} - Y_1}{\rho_{es\_1}} \cdot \Delta y$$
$$+ \frac{Z_{es} - Z_1}{\rho_{es\_1}} \cdot \Delta z + c\delta t_i^1 + \delta I_i^1(t) + \delta T_i^1(t) \qquad (1.25)$$

$$\tilde{\rho}_i^2 = \rho_{es\_2} + \frac{X_{es} - X_2}{\rho_{es\_2}} \cdot \Delta x + \frac{Y_{es} - Y_2}{\rho_{es\_2}} \cdot \Delta y +$$
$$\frac{Z_{es} - Z_2}{\rho_{es\_2}} \cdot \Delta z + c\delta t_i^2 + \delta I_i^2(t) + \delta T_i^2(t) \qquad (1.26)$$

$$\tilde{\rho}_i^2 = \rho_{es\_2} + \frac{X_{es} - X_2}{\rho_{es\_2}} \cdot \Delta x + \frac{Y_{es} - Y_2}{\rho_{es\_2}} \cdot \Delta y +$$
$$\frac{Z_{es} - Z_2}{\rho_{es\_2}} \cdot \Delta z + c\delta t_i^2 + \delta I_i^2(t) + \delta T_i^2(t) \qquad (1.27)$$

$$\tilde{\rho}_i^4 = \rho_{es\_4} + \frac{X_{es} - X_4}{\rho_{es\_4}} \cdot \Delta x + \frac{Y_{es} - Y_4}{\rho_{es\_4}} \cdot \Delta y$$
$$+ \frac{Z_{es} - Z_4}{\rho_{es\_4}} \cdot \Delta z + c\delta t_i^4 + \delta I_i^4(t) + \delta T_i^4(t) \qquad (1.28)$$

Equations (1.25)–(1.28) is the linearized form of the code measurement pseudo range observation equation, which is written as the general form, i.e.

$$\tilde{\rho}_i^j(t) = (\rho_i^j(t))_0 - k_i^j(t)\delta X_i - l_i^j(t)\delta Y_i - m_i^j(t)\delta Z_i + c\delta t_i^j + \delta I_i^j(t) + \delta T_i^j(t)$$

$$(1.29)$$

where, k, l, m are the direction cosine values from the receiver to the satellite.

In order to find out the position of the user and the clock difference between the satellite and the receiver, it is only necessary to calculate $\Delta x, \Delta y, \Delta z, \Delta t_i^j$. Because the user observes four satellites at the same time, the clock difference between the four satellites and the receiver is the same. It is recorded as Δt. Then Eqs. (1.25)–(1.28) can be written as follows:

$$
\begin{bmatrix} \tilde{\rho}_i^1 - \rho_{es\_1} \\ \tilde{\rho}_i^2 - \rho_{es\_2} \\ \tilde{\rho}_i^3 - \rho_{es\_3} \\ \tilde{\rho}_i^4 - \rho_{es\_4} \end{bmatrix} = \begin{bmatrix} \frac{X_{es}-X_1}{\rho_{es\_1}} & \frac{Y_{es}-Y_1}{\rho_{es\_1}} & \frac{Z_{es}-Z_1}{\rho_{es\_1}} & c \\ \frac{X_{es}-X_2}{\rho_{es\_2}} & \frac{Y_{es}-Y_2}{\rho_{es\_2}} & \frac{Z_{es}-Z_2}{\rho_{es\_2}} & c \\ \frac{X_{es}-X_3}{\rho_{es\_3}} & \frac{Y_{es}-Y_3}{\rho_{es\_3}} & \frac{Z_{es}-Z_3}{\rho_{es\_3}} & c \\ \frac{X_{es}-X_4}{\rho_{es\_4}} & \frac{Y_{es}-Y_4}{\rho_{es\_4}} & \frac{Z_{es}-Z_4}{\rho_{es\_4}} & c \end{bmatrix} \cdot \begin{bmatrix} \Delta x \\ \Delta y \\ \Delta z \\ \Delta t \end{bmatrix} + \begin{bmatrix} \delta I_i^1 + \delta T_i^1(t) \\ \delta I_i^2 + \delta T_i^2(t) \\ \delta I_i^3 + \delta T_i^3(t) \\ \delta I_i^4 + \delta T_i^4(t) \end{bmatrix}
$$

$$(1.30)$$

If the last term of Eq. (1.30) is ignored, that is, the refraction correction term of ionosphere and troposphere, then

$$
\begin{bmatrix} \Delta x \\ \Delta y \\ \Delta z \\ \Delta t \end{bmatrix} = \begin{bmatrix} \frac{X_{es}-X_1}{\rho_{es\_1}} & \frac{Y_{es}-Y_1}{\rho_{es\_1}} & \frac{Z_{es}-Z_1}{\rho_{es\_1}} & c \\ \frac{X_{es}-X_2}{\rho_{es\_2}} & \frac{Y_{es}-Y_2}{\rho_{es\_2}} & \frac{Z_{es}-Z_2}{\rho_{es\_2}} & c \\ \frac{X_{es}-X_3}{\rho_{es\_3}} & \frac{Y_{es}-Y_3}{\rho_{es\_3}} & \frac{Z_{es}-Z_3}{\rho_{es\_3}} & c \\ \frac{X_{es}-X_4}{\rho_{es\_4}} & \frac{Y_{es}-Y_4}{\rho_{es\_4}} & \frac{Z_{es}-Z_4}{\rho_{es\_4}} & c \end{bmatrix}^{-1} \cdot \begin{bmatrix} \tilde{\rho}_i^1 - \rho_{es\_1} \\ \tilde{\rho}_i^2 - \rho_{es\_2} \\ \tilde{\rho}_i^3 - \rho_{es\_3} \\ \tilde{\rho}_i^4 - \rho_{es\_4} \end{bmatrix}
$$

$$(1.31)$$

In this way, Δx, Δy, Δz, and Δt are obtained. For this kind of approximate calculation, considering that the accuracy of approximate coordinates is relatively low and the value of coordinate correction (Δx, Δy, Δz) is relatively large, the coordinates ($X_{es} + \Delta x$, $Y_{es} + \Delta x$, $Z_{es} + \Delta x$) are used to replace the initial user's approximate position coordinates (X_{es}, Y_{es}, Z_{es}). Repeating the above calculation, and iterating until there is no obvious difference between the two iterations, and finally the user's coordinates (X, Y, Z) are obtained.

1.3.3 Carrier Phase Measurement

It is difficult to meet the requirements of precision measurement because of the low accuracy of satellite ranging code. Therefore, when the measurement accuracy is required to reach centimeter level or even millimeter level, carrier phase measurement technology must be used.

1.   The principle of carrier phase measurement

The receiver needs to measure the phase of the carrier signal at the satellite and the receiver at the same time, and then make the difference between them to obtain the pseudo range from the satellite to the receiver. For example, at a certain time, the satellite S sends a carrier signal, the phase of the signal at the satellite S is φ_S, and the phase at the receiver R is φ_R. φ_S and φ_R are the carrier phase including the number of whole cycles calculated from a certain point. For the convenience of calculation, the cycle is taken as the unit, one cycle corresponds to 360° phase change, and one carrier wavelength in the distance. If the wavelength of the carrier is λ, the distance between the satellite S and the receiver R is

$$\rho = \lambda(\varphi_S - \varphi_R) \tag{1.32}$$

But in practical work, it is impossible to measure φ_S. Instead, the oscillator of the receiver generates a reference signal with the same frequency and initial phase as the satellite signal, so that at any instant, the phase of the reference signal of the receiver is equal to the signal phase of the satellite S.

In the actual carrier phase measurement, the measured phase difference includes the whole cycle part and the fractional part less than one whole cycle. The phase observation value is

$$\varphi = \varphi_S - \varphi_R = N + \Delta\varphi \tag{1.33}$$

where N is the number of integer cycle and $\Delta\varphi$ is the fraction of less than one whole cycle. However, because the carrier wave is a simple sine wave, it does not have any recognizable identification. Therefore, it is impossible to know exactly which period of phase is being measured. In other words, this unknown integer N is called integer ambiguity.

2. Carrier phase observation equation

Carrier phase observation is a function of the position of the receiver and satellite. Only when the functional relationship between them is obtained, the position of the receiver can be calculated from the observation.

Assuming that the carrier signal is a sine wave $y = A \sin(\omega t + \varphi_0)$, the time when the satellite transmits the carrier signal is t^j, if the clock of the receiver has no error, then the time when the receiver generates the duplicate signal is also t^j, and the time when the receiver receives the satellite signal is t_k, then the time when the carrier signal propagates is $t_k - t^j = \Delta t + N \cdot T$. We can get the distance between the stations is

$$\rho = c(\Delta t + N \cdot T) = c\frac{\Delta\varphi' + N \cdot 2\pi}{2\pi f} = \lambda\frac{\Delta\varphi'}{2\pi} + N \cdot \lambda = \lambda \cdot \Delta\varphi + N \cdot \lambda \tag{1.34}$$

where, $\Delta\varphi'$ in radians, and N and $\Delta\varphi$ in cycles. From Eq. (1.34), it can be concluded that

$$\lambda \cdot \Delta\varphi = \rho - N \cdot \lambda \tag{1.35}$$

When the receiver locks the satellite signal at the moment t_0 and starts to measure, it can only measure the fractional part $\Delta\varphi$ of the phase less than one cycle. That is, the left end of Eq. (1.35) is measurable, while the number of phase cycles N at the initial time is unknown. The two terms at the right end of Eq. (1.35) are unknown. As long as the satellite does not lose its lock, till the moment t_i, the phase difference between the satellite and the receiver will contain three items: first, the whole cycle part of the initial time, which is a fixed value; second, the whole cycle change part, which can be measured by the whole wave counter; third, the fractional part which is less than the whole cycle.

The sum of the whole cycle variation part and the insufficient whole cycle part is expressed as $\varphi(t_i)$. The influence of the receiver clock error is considered. The carrier phase observation equation is

$$\lambda \cdot \varphi(t_i) = \rho + c \cdot \delta t(t_i) - N \cdot \lambda \tag{1.36}$$

After substituting the coordinates of satellite and receiver into Eq. (1.36), and considering the ionosphere and troposphere correction, we can get

$$\lambda \cdot \varphi(t_i) = a_k^j \delta X + b_k^j \delta Y + c_k^j \delta Z + c \cdot \delta t(t_i) - N \cdot \lambda + l_0 \tag{1.37}$$

where the geometric distance approximation and ionospheric and tropospheric corrections are included in l_0. There are five unknowns in Eq. (1.37). If five satellites are observed, there will be nine unknowns.

The pseudo range equation of approximate carrier phase observation

$$\tilde{\rho}_i^j(t) = \rho_i^j(t) + \delta I_i^j(t) + \delta T_i^j(t) + c\delta t_i - c\delta t^j - \lambda N_i^j(t_0) \tag{1.38}$$

After linearization

$$\tilde{\rho}_i^j(t) = (\rho_i^j(t))_0 - (k_i^j(t)\delta X_i + l_i^j(t)\delta Y_i + m_i^j(t)\delta Z_i) + \\ \delta I_i^j(t) + \delta T_i^j(t) + c\delta t_i - c\delta t^j - \lambda N_i^j(t_0) \tag{1.39}$$

3.   Main problems of carrier phase observation

In carrier phase measurement, it is impossible to directly measure the integer number of the phase change of the satellite carrier signal on the propagation path, which leads to the integer ambiguity problem. In addition, in the process of tracking GNSS satellite for observation, the satellite signal is often loss of lock due to the interference of external noise signal and the shielding of receiver antenna, resulting in cycle slips. Integer ambiguity resolution is a key and difficult problem in carrier phase observation, which can be solved by appropriate data processing methods.

If the dynamic absolute positioning of phase measurement pseudo range is to be carried out, the receiver should be fixed at a point for observation for a period of time before the observation to obtain the integer ambiguity. This process is called initialization, and then the dynamic absolute positioning of phase measurement pseudo range can be carried out.

When observing the carrier phase, it should be noted that the change part of the whole cycle number is recorded by the counter, and the signal cannot be interrupted during this period. If the signal arriving at the receiver during this period is blocked, resulting in the loss of lock, the whole cycle counting will be suspended during the blocking period. The counting will continue after the blocking is removed, which will lose some phase cycles during the blocking period. This situation is called cycle slip. Another reason for cycle slip is strong electromagnetic interference.

1.3.4 Accuracy of Satellite Navigation and Positioning

The positioning accuracy of GNSS mainly depends on two factors: the geometric distribution of satellites and the measurement error. GNSS positioning error can be expressed as the product of geometric precision factor GDOP (Geometric Dilution of Precision) and total equivalent distance error σ. This section will discuss how to measure the source of error, how to evaluate the positioning accuracy, and how the geometric distribution of satellites affects the positioning accuracy.

1.   GNSS measurement error

GNSS satellite positioning is to determine the three-dimensional coordinates of a point on the ground by receiving the carrier phase, pseudo range and ephemeris data transmitted by the satellite on the ground. The main sources of GNSS measurement error are GNSS satellite, signal propagation process and receiver. In the process of high-precision measurement, the positioning accuracy will also be affected by the load tide, solid tide and relativistic effect related to the overall movement of the earth.

1)   Errors related to GNSS satellites (Space segment errors)

These errors mainly include satellite ephemeris error and satellite clock error, which are caused by the ground monitoring part of GNSS cannot accurately predict and measure the clock drift of satellite clock and the orbit of satellite.

Although high-precision atomic clocks are used on satellites, they still inevitably have errors. This kind of error includes not only systematic error (error caused by frequency offset, clock error, frequency drift, etc.), but also random error. The systematic error is larger than the random error, but it can be corrected by the model. Therefore, the random error becomes an important symbol to measure the quality of satellite clock.

In the GNSS ground monitoring part, ephemeris parameters are used to describe and predict the orbit of the satellite. But the GNSS satellite is bound to be affected by various complex perturbations during its operation. There must be differences between the predicted orbit model and the real orbit of the satellite. The ephemeris errors of each satellite are generally independent of each other.

2)   Error related to signal propagation (Environmental segment error)

The GNSS signal needs to pass through the atmosphere when it propagates from the satellite to the receiver, and the influence of the atmosphere on the signal propagation is mainly the atmospheric delay. Atmospheric time delay is usually divided into tropospheric delay and ionospheric delay.

The atmosphere 50–1000 km above the ground is called the ionosphere. The atmospheric molecules and atoms in the ionosphere will decompose into electrons and atmospheric ions under the irradiation of sunlight and high-energy extraterrestrial rays. When the electromagnetic wave passes through the ionosphere filled with electrons, its propagation speed and direction will change, resulting in the ionospheric deviation error of GNSS measurement results.

The troposphere is located at the bottom of the atmosphere, and its top is about 40 km above the ground. The troposphere concentrates 99% of the mass of the atmosphere. Nitrogen, oxygen and water vapor are the main reasons for the delay of GNSS signal propagation. When the satellite signal passes through the troposphere, the propagation speed will change, which will cause the corresponding error of the measurement results. The error will be affected by the pressure, temperature and other factors.

In addition to receiving the electromagnetic wave signal transmitted from GNSS satellite through a straight line, the receiver antenna may also receive one or more signals reflected once or more by the electromagnetic wave through the surrounding ground objects, which is called multipath effect. Multipath effect can also produce errors in GNSS measurement results, which are affected by the performance of the receiver antenna and the environment around the receiver.

3) Receiver related error (User segment error)

This part has a wide range of meanings, including the position error of the receiver (the position of the zero phase center of the receiver antenna does not coincide with the position of the receiver), the clock error of the receiver, the thermal noise of the electronic components of each part, the signal quantization error, the algorithm error between the measured code phase and the carrier phase, and the calculation error in the receiver software.

2. Precision factor

In navigation, generally, the precision factor DOP (Dilution of Precision) is used to evaluate the positioning results. The precision factor is also called precision coefficient or error coefficient. Its influence on the positioning results is as follows

$$m_x = \text{DOP} \cdot \sigma \tag{1.40}$$

DOP is the function of the main diagonal elements in the weight coefficient matrix in the pseudo range absolute positioning

$$\mathbf{Q}_x = (\mathbf{A}_i^{\mathrm{T}} \mathbf{A}_i)^{-1} \tag{1.41}$$

Or expressed as

$$\mathbf{Q}_x = \begin{bmatrix} q_{11} & q_{12} & q_{13} & q_{14} \\ q_{21} & q_{22} & q_{23} & q_{24} \\ q_{31} & q_{32} & q_{33} & q_{34} \\ q_{41} & q_{42} & q_{43} & q_{44} \end{bmatrix} \tag{1.42}$$

The elements in Eq. (1.42) reflect the positioning precision and spatial correlation information of different parameters under certain geometric distribution, which is the basis for evaluating the positioning results. Using different combinations of these elements, the positioning accuracy can be evaluated from different aspects.

The weight coefficient matrix of Eq. (1.42) is usually given in the space rectangular coordinate system. But in order to estimate the position precision of the observation station, its expression in the geodetic coordinate system is often used. Suppose that in the geodetic coordinate system, the weight coefficient matrix of the corresponding point is

$$\mathbf{Q}_B = \begin{bmatrix} q_{11} & q_{12} & q_{13} \\ q_{21} & q_{22} & q_{23} \\ q_{31} & q_{32} & q_{33} \end{bmatrix} \tag{1.43}$$

According to the propagation law of variance and covariance

$$\mathbf{Q}_B = \mathbf{H} \mathbf{Q}'_x \mathbf{H}^\mathrm{T} \tag{1.44}$$

where,

$$\mathbf{Q}'_x = \begin{bmatrix} q_{11} & q_{12} & q_{13} \\ q_{21} & q_{22} & q_{23} \\ q_{31} & q_{32} & q_{33} \end{bmatrix} \tag{1.45}$$

$$\mathbf{H} = \begin{bmatrix} -\sin B \cos L & -\sin B \sin L & \cos B \\ -\sin L & \cos L & 0 \\ \cos B \cos L & \cos B \sin L & \sin B \end{bmatrix} \tag{1.46}$$

In practice, according to different requirements, different precision evaluation models and corresponding precision factors can be selected, usually including the following.

(1) Three dimensional position precision factor PDOP (Position DOP),

$$\mathrm{PDOP} = (q_{11} + q_{22} + q_{33})^{1/2} \tag{1.47}$$

The corresponding three-dimensional positioning accuracy is

$$m_P = \text{PDOP} \cdot \sigma \tag{1.48}$$

(2)   Horizontal component precision factor HDOP (Horizontal DOP),

$$\text{HDOP} = (q_{11} + q_{22})^{1/2} \tag{1.49}$$

The accuracy of the corresponding horizontal component is

$$m_H = \text{HDOP} \cdot \sigma \tag{1.50}$$

(3)   Vertical component precision factor VDOP (Vertical DOP),

$$\text{VDOP} = (q_{33})^{1/2} \tag{1.51}$$

The precision of the corresponding vertical component is

$$m_V = \text{VDOP} \cdot \sigma \tag{1.52}$$

(4)   Receiver clock error precision factor TDOP (Time DOP),

$$\text{TDOP} = (q_{44})^{1/2} \tag{1.53}$$

The corresponding clock error precision is

$$m_T = \text{TDOP} \cdot \sigma \tag{1.54}$$

(5)   At the same time, there is a geometric precision factor GDOP (Geometric DOP). The geometric precision factor is an accuracy factor that integrates PDOP and TDOP to describe the comprehensive influence of three-dimensional position and time errors,

$$\text{GDOP} = \left(\text{PDOP}^2 + \text{TDOP}^2\right)^{1/2} = (q_{11} + q_{22} + q_{33} + q_{44})^{1/2} \tag{1.55}$$

The corresponding space–time precision is

$$m_G = \text{GDOP} \cdot \sigma \tag{1.56}$$

3.   The geometric distribution of satellites

The precision factor affects the absolute positioning error of GNSS, and the geometric distribution of the measured satellites affects the DOP. Because the choice of observation satellites and the movement of satellites are different, the geometric distribution of the measured satellites in space is constantly changing. The value of precision factor is also constantly changing.

Since the geometric distribution of satellites affects the DOP, it is a question of concern which one is more suitable. Theoretical analysis shows that: assuming that the observation station and four satellites form a hexahedron, the GDOP is proportional to the reciprocal of the volume of the hexahedron V, that is

$$GDOP \propto \frac{1}{V} \tag{1.57}$$

It can be seen from Eq. (1.57) that the larger the spatial distribution range of the satellite, the larger the volume of the hexahedron, and the smaller the GDOP value. On the contrary, the smaller the volume of the hexahedron, the larger the GDOP value.

Theoretical analysis shows that the volume of hexahedron is the largest when the angle between any two directions is close to $109.5°$ from the observation station to the four satellites. However, in order to weaken the influence of atmospheric refraction, the altitude angle of the satellite should not be too low. Therefore, the volume of hexahedron must be as close to the maximum as possible on the premise of meeting the requirements of satellite altitude angle. It is generally believed that when the altitude angle meets the above conditions, one satellite is at the zenith and the other three satellites are about $120°$ apart, the volume of hexahedron is close to the maximum, which can be used as a reference for selecting and evaluating satellite distribution in practical work. Figure 1.14 shows the comparison of GDOP.

1.4 Error Analysis of Satellite Navigation System

1.4.1 Introduction of Satellite Navigation System Error

According to the causes and properties of the measurement errors of satellite navigation and positioning, the errors of satellite navigation system can be divided into systematic errors (also known as bias) and accidental errors. Systematic error is a error which has a great influence on the satellite navigation system, and the maximum is several hundred meters. Systematic error is usually related to some variables such as time, position and atmosphere. Therefore, the influence of systematic error can be eliminated or suppressed by modeling the source of systematic error. Accidental errors include random noise, observation error, multipath effect and other external influence errors with random characteristics in satellite signal generating part and receiver signal receiving and processing part. Accidental errors are random and have

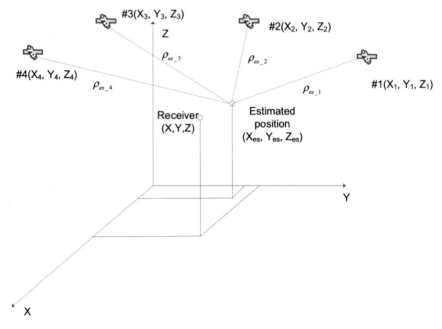

Fig. 1.13   Pseudo range positioning method

little influence on the satellite positioning system, usually in the millimeter to meter level.

All kinds of errors in satellite navigation and positioning can be divided into three categories according to the stage of error source generation: one is the error generated in the space segment, the other is the error generated in the environment segment, and the third is the error generated in the user segment.

(1)   The errors in the space segment are mainly related to the satellite itself, including the errors caused by satellite orbit error, satellite clock error, earth rotation and relativistic effect. The satellite orbit parameters and clock model are given by the navigation message broadcast by the satellite. But in fact the satellite is not exactly in the position predicted by the navigation message. Even if the satellite clock is corrected by the satellite clock model in the navigation message, it will not be synchronized with the satellite navigation system. These errors are uncorrelated among satellites, and they have the same effect on the pseudo range and carrier phase measurements. The errors of satellite orbit and satellite clock in space segment are related to the position and number of ground tracking stations, the standard time of satellite navigation system, the model describing satellite orbit and the geometric structure of satellite in space.

(2)   The errors generated in the environmental phase include the errors related to satellite signal transmission path and observation method, such as ionospheric and tropospheric delay, multipath effect and so on. When the satellite signal

passes through the atmosphere, it will refract in the ionosphere and tropo-sphere, resulting in signal transmission error. High buildings on the ground and water surface will also reflect satellite signals, resulting in multipath effect and interference error.

(3) The errors in the user segment are mainly caused by the clock deviation of the receiver and the phase center deviation of the antenna. Because the speed of satellite electromagnetic wave signal transmission is the speed of light, the clock deviation of the receiver will have a great impact on the results of satellite navigation and positioning, which is usually set as an unknown value to solve. In precision measurement, the phase center of the antenna itself is not consistent with the physical center of the actual measurement, which will also bring errors to the measurement.

The errors caused by various reasons have quite complex spectrum characteristics and other characteristics, and some error sources may still be related. Therefore, the complex cross coupling relationship must be analyzed in precision measurement. In this book, in order to make readers more clearly understand the generation and elimination methods of various error sources, it is assumed that the error sources are uncorrelated, and different equations are used to describe their characteristics.

1.4.2 Space Segment Error

1. Satellite ephemeris error

Satellite ephemeris error, also known as satellite orbit error, is the difference between the satellite position calculated by ephemeris parameters or other orbit information and the actual position of the satellite.

The position of the ground monitoring station of the satellite navigation system is known accurately. There are accurate atomic clocks in the station. In orbit deter-mination, several monitoring stations distributed in different areas can track and monitor the same satellite, measure the distance between stations, and then determine the position of the satellite according to the observation equation. The positioning method of solving the satellite position from the known ground monitoring position is called reverse ranging positioning (or orbit determination). The main control station performs the best filtering processing on the long-term measured data of the moni-toring station, forms the ephemeris, uploads it into the satellite, and then transmits it to the user in the form of navigation message.

Because satellites are affected by many kinds of celestial perturbations, it is diffi-cult for ground monitoring stations to fully and reliably measure the effects of these perturbations. Therefore, the orbit of satellites measured contains errors. At the same time, the quality of the monitoring system, such as the number and spatial distribution of the tracking stations, the number and accuracy of the orbit parameters, the orbit model used to calculate the orbit and the perfection of the planning software, will also lead to ephemeris error. In addition, the satellite ephemeris obtained by the user is not real-time, but calculated by the ephemeris parameters corresponding to a certain time in the navigation message received by the user, which will also lead to errors in the calculation of satellite position. The influence of broadcast ephemeris error on the coordinates of observation station can reach several meters or even hundreds of meters. In addition, ephemeris error is a systematic error, which cannot be eliminated by repeated observation. Therefore, we must consider the errors and eliminate them when we revise the ephemeris model or when we receive and settle accounts.

2.   Satellite clock error

Satellite clock error refers to the non-synchronous deviation between satellite clock and navigation system standard time. Although high-precision atomic clocks (such as cesium and rubidium clocks) are used on satellites, there will be clock error, frequency offset, frequency drift and random error between these clocks and the standard time of satellite navigation system, and these frequency offset and frequency drift will change with time. Since the satellite position is a function of time, the observation quantity of satellite navigation system is based on the precise measurement time. The satellite clock error will affect the results of PRN code ranging and carrier phase measurement. The total amount of this error can reach 1 ms, and the equivalent distance error can reach 300 km.

Satellite navigation system is essentially a time/range measurement positioning system. The positioning accuracy of satellite navigation system is closely related to the clock error. Taking GPS as an example, the unified time standard of GPS measurement is GPS time system, which is determined and maintained by GPS ground monitoring system. All GPS satellites are equipped with high-precision atomic clocks to ensure the high-precision of satellite clocks. However, there is still a total deviation and drift of 0.1 to 1 ms between them and GPS standard time. The resulting equivalent distance error will reach 30 to 300 km, which must be accurately corrected.

3.   The influence of relativistic effect

Relativistic effect is a phenomenon of relative clock difference between satellite clock and receiver clock caused by different states, including special relativistic effect and general relativistic effect.

According to the special theory of relativity, an oscillator with a frequency of f is installed on the carrier with a flight speed of v. Because the carrier is moving, the clock will have a frequency change for the ground observer. Because of the expansion of time, the frequency of the clock will change with the change of speed. Under the influence of special relativity, the clock will slow down when it is installed on the satellite.

In addition, according to the general theory of relativity, the frequency of oscillators on different equipotential surfaces will change due to the different gravitational potential. This phenomenon is often called gravitational frequency offset. According to general relativity, the frequency of a clock is related to its gravitational potential. Under the action of general relativity, the frequency of the satellite clock will be faster.

Under the combined effect of special relativity and general relativity, the satellite clock can walk faster than the clock installed on the ground. In order to eliminate the influence of relativistic effect, the clock on the satellite should be adjusted slower than that on the ground. However, due to the changes of earth motion, satellite orbit height and earth gravity field, the relativistic effect mentioned above is not constant and cannot be ignored for precise positioning.

1.4.3 Environmental Segment Error

The errors in the environment segment mainly include ionospheric delay error, tropospheric delay error, multipath effect error and other disturbances. Atmospheric refraction effect refers to when the signal passing through the atmosphere, the speed will change, the propagation path will also bend, which is also known as atmospheric delay. In the measurement and positioning of GNSS, only the change of signal propagation speed is usually considered. In dispersive medium, the refraction effect of different frequency signals is different. In non-dispersive medium, the refraction effect of different frequency signals is the same. For the signal of GNSS, the ionosphere is a dispersive medium and the troposphere is a non-dispersive medium.

1. Ionospheric delay error

The atmosphere can be divided into ionosphere and troposphere. The ionosphere is 50–1000 km above the ground. The ionosphere is mainly composed of gas ionized by solar radiation and contains a large number of free electrons and positive ions. Therefore, when the electromagnetic wave of the satellite signal passes through the ionosphere, the propagation speed and path of the signal will change due to the different charge density. If the propagation time is still multiplied by the speed of light in vacuum to calculate the propagation distance of the signal, a large error will be caused, which is the ionospheric delay error.

The ionospheric delay error of satellite signal has the following characteristics.

(1) For the same observation station, the ionospheric delay errors in different directions are different. The ionospheric delay error in the zenith direction of the observation station is the smallest. The lower the satellite elevation, the greater the error caused by the ionospheric delay.

(2) The ionospheric delay errors of satellite signals observed at different epoch are different at the same observation station. The ionospheric delay error in daytime is larger than that in night.

(3) The ionospheric delay error is different at different location. However, the ionospheric delay error has a strong geographical correlation. For the same satellite, the ionospheric delay error of the signal received by the observation station not far away (within 50 km) is basically the same.

The range residual of ionospheric delay error corrected by dual frequency measurement is centimeter level. Therefore, dual frequency receiver is generally used in precision measurement. It is impossible to measure the ionospheric delay using single frequency receivers. In order to reduce the influence of ionospheric delay error, the measured ionospheric model provided by navigation message in satellite signal or the local ionospheric statistical model are used to correct the observation. However, due to the large variation of the number of ionospheric electrons, the correction effect of the measured model provided in the navigation message is better than that of the historical data statistical model.

One method is to use two or more receivers to observe the same satellite or the same group of satellites synchronously, and then calculate the difference value of the synchronous observation value to weaken the influence of ionospheric refraction. Especially when the distance between two or more GNSS stations is relatively close (about 20 km). Because the paths of satellite signals arriving at different stations are similar and the ionospheric medium conditions are similar, the influence of ionospheric refraction can be significantly reduced by calculating the difference of the same satellite synchronous observations from different stations. For single frequency receiver, the effect of this method is particularly obvious.

2.   Tropospheric delay error

The troposphere is located in the bottom layer of the atmosphere within 40 km above the ground, accounting for 99% of the total atmospheric mass. When the troposphere contacts with the ground, the radiant heat energy is obtained from the ground. The temperature will decrease about 6.5 °C for every 1 km increase in the vertical direction, while the temperature difference in the horizontal direction (north–south direction) will not exceed 1 °C for every 100 km. The troposphere has strong convection, in which wind, rain, cloud, fog, snow and other weather phenomena appear. The atmosphere in this layer contains not only various gas elements, but also water droplets, ice crystals, dust and other impurities, which have a great impact on electromagnetic wave propagation. In the troposphere, due to the existence of refraction, the propagation velocity of electromagnetic wave will change.

The atmospheric density in the troposphere is higher than that in the ionosphere, and the atmospheric state is more complex. Therefore, when the satellite signal passes through the troposphere, the path will also bend. Besides height variation, tropospheric refractive index is closely related to atmospheric pressure, temperature and humidity. Due to the strong convective effect of atmosphere, the changes of atmospheric pressure, temperature, humidity and other factors are very complex. It is difficult to accurately model the changes of atmospheric tropospheric refractive index and its influence at present, and there are many correction models of tropospheric delay based on empirical values. Even with the real-time meteorological data, the residual error of the propagation path delay of electromagnetic wave after tropospheric refraction correction is still about 5% of the tropospheric effect.

3.   Multipath effect

Multipath effect, also known as multipath error, refers to the fact that when a satellite transmits signals to the ground, the receiver may receive one or more reflected satellite

Fig. 1.14 GDOP comparison

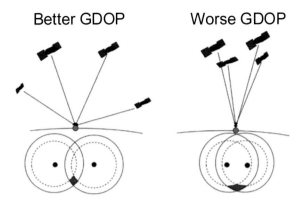

signals from surrounding buildings and water surface in addition to the direct signals from the satellite. The superposition of these signals will cause the change of the position of the measurement reference point (the phase center of the antenna of the satellite navigation receiver), thus resulting error in the observation results.

Multipath effect is mainly affected by the reflection surface near the receiver, such as tall buildings, high-rise structures of warships, outer surfaces of space shuttles or other space vehicles, as shown in Fig. 1.15. In Fig. 1.15, the satellite signal arrives at the receiver antenna through three different paths, one of which arrives directly and the other two arrive indirectly. Therefore, the signal received by the receiver antenna has relative phase offset, and these phase differences are proportional to the path length. Since the path shape of the reflected signal is arbitrary, there is no general model for multipath effect. However, the influence of multipath effect can be estimated by the measurement difference of multiple carrier phase. The principle is that troposphere, clock error and relativistic effect affect the measurement of PRN code and carrier phase by the same amount, and ionosphere and multipath effect are frequency dependent. Therefore, once the ionospheric independent PRN code distance and carrier phase are obtained, all the effects mentioned above can be eliminated except for multipath, and the remaining effects are mainly multipath.

1.4.4 User Segment Error

User segment error mainly refers to the relevant errors generated on the user receiving equipment, including observation error, receiver clock error, receiver antenna phase center deviation, carrier phase observation of the cycle slip and earth rotation and tide phenomenon on the receiver.

1. Observation error

The observation error is not only related to the observation resolution of the software and hardware of the receiver, but also related to the installation accuracy of the antenna. According to the experimental results, it is generally considered that the resolution error of the observation is 1% of the signal wavelength. The observation accuracy of the code signal and carrier signal of the satellite navigation system, taking the GPS system as an example, is shown in Table 1.8.

The observation error caused by antenna installation accuracy refers to the antenna alignment error, antenna leveling error and the error of measuring antenna phase center height (antenna height). For example, when the height of the antenna is 1.6 m, if the leveling error of the antenna is $0.1°$, the alignment error of the optical centering device is about 3 mm. Therefore, attention should be paid to leveling the antenna and careful alignment in order to reduce the installation error.

2.   Receiver clock error

The receiver of generally adopts high-precision quartz clock, and its daily frequency stability is about 10^{-11}. If the synchronization error between station clock and satellite clock is $1\mu s$, the equivalent distance error is about 300 m. In order to further improve the accuracy of the station clock, the constant temperature crystal oscillator can be used. But its volume and power consumption are large, and the frequency stability can only be improved by 1–2 orders of magnitude. In single point positioning, the clock error is usually taken as an unknown parameter and solved together with the position parameter of the observation station. In positioning, if each observation instant clock error is assumed to be independent, the processing is relatively simple. Therefore, this method is widely used in dynamic absolute positioning. In the process of carrier phase relative positioning, the method of single difference between satellites and double difference between stations can effectively eliminate the clock error of receiver. When the positioning accuracy is high, the method of external frequency standard (time standard) can be used, such as rubidium atomic clock or cesium atomic clock. This method is often used in fixed observation.

3.   Phase center deviation of receiver antenna

The position deviation of the receiver refers to the deviation of the phase center of the receiver antenna from the geometric center of the antenna. In the positioning process of GNSS, whether it is code pseudo range or phase pseudo range, the observed value is the measured distance from the satellite to the antenna phase center. The antenna alignment is based on the geometric center of the antenna. Therefore, the requirement for the antenna is that its phase center and geometric center should be consistent as far as possible.

Table 1.8 Observation error caused by observation resolution

Signals	wavelength/m	Observation error/m
C/A code	293	2.9
P code	29.3	0.3
L1 carrier	0.1905	2.0×10^{-5}
L2 carrier	0.2445	2.5×10^{-5}

In fact, the position of the phase center of the antenna will change with the intensity and direction of the signal input. Therefore, the instantaneous position of the phase center (called apparent phase center) during observation will be different from the theoretical position of the phase center. The difference between the antenna phase center and the geometric center is called the deviation of the antenna phase center. This deviation will cause positioning error. According to the performance of the antenna, it can reach tens of millimeters or several centimeters. Therefore, for the precision relative positioning, this influence cannot be ignored.

How to reduce the offset of phase center is a key problem in antenna design. In practical measurement, if the same type of antenna is used to observe the same group of satellites at two or more receiver not far away, the effect of phase center deviation can be weakened by calculating the difference of observation values. However, at this time, the antenna of each observation station should be oriented according to the bearing mark attached to the antenna disk to meet certain accuracy requirements. In addition, the phase center deviation correction of satellite and receiver antenna should be considered when establishing the observation equation. The phase center deviation can be corrected by correcting the coordinates of the satellite or the receiver, or by directly correcting the observations.

4.   The cycle slip of carrier phase observation

At present, the carrier phase observation method is widely used, which can improve the positioning accuracy to millimeter level. However, in the observation epoch t, the receiver can only provide the fractional part of the carrier phase and the integer number of the carrier phase change from the lock epoch t_0 to the observation epoch t. It cannot directly obtain the integer number of the carrier phase change in the whole propagation path at the lock epoch. The principle is shown in Fig. 1.16.

Therefore, the integer ambiguity is needed in the observation of carrier phase, and the accuracy of the calculated value will affect the ranging accuracy. It is an important work to determine the integer ambiguity N_0 in carrier phase measurement. Because

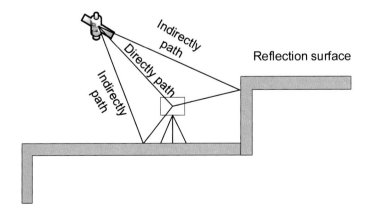

Fig. 1.15 Multipath effect

the satellite navigation receiver can process pseudo range measurement and carrier phase measurement simultaneously, and $\lambda \cdot N_0$ can be obtained by the pseudo range observation subtracting the actual observation value of carrier phase measurement (converted into distance unit). However, due to the low accuracy of pseudo range measurement, it is necessary to use more average values of $\lambda \cdot N_0$ to get the accurate number of integer cycles. However, the accuracy of the above methods is low. In actual measurement, the integer solution or real solution of integer ambiguity is usually obtained according to the length of baseline. Because the integer ambiguity should be an integer in theory, using this property can improve the accuracy of the solution. This method is generally used in short baseline positioning. When the baseline is long, the correlation of errors will be reduced, and many errors are not eliminated perfectly. Therefore, neither the baseline vector nor the integer ambiguities can be estimated accurately. In this case, the real solution is usually taken as the final solution.

If the receiver can keep the continuous tracking of the satellite signal in the observation process, the integer ambiguity N_0 will remain unchanged, and the integer count $Int(\varphi)$ will also remain continuous. However, when the receiver is unable to continuously track the satellite signal for some reason, the N_0 will change after it is locked again, and $Int(\varphi)$ will not keep continuous with the previous value. This phenomenon is called cycle slip. The diagram is shown in Fig. 1.17.

When the carrier phase method is used in ranging, in addition to the calculation of integer ambiguity, cycle slip may also occur in the observation process. It is worth noting that the cycle slip phenomenon is very easy to occur in carrier phase measurement, which has a great impact on the phase pseudo range observations, and is a very important problem in precise positioning data processing.

The cycle slip phenomenon will destroy the regular variation of carrier phase measurements $Int(\varphi) + \Delta\varphi$ with time. The radial velocity of the satellite is very

Fig. 1.16 Principle of carrier phase observation

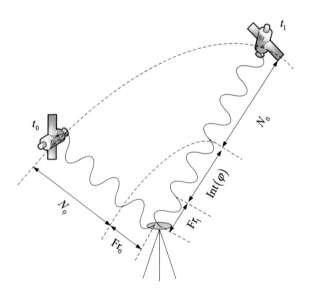

Fig. 1.17 Cycle slip

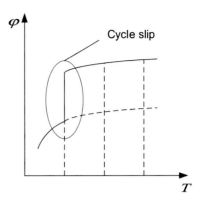

large, up to 0.9 km/s, and the integer cycle count can change thousands of cycles per second. Therefore, the difference between adjacent observations is also large. If the cycle slip is only a few weeks or dozens of weeks, it is not easy to be found. In this case, the cycle slip can be detected by finding multiple differences between two adjacent observations.

5.   Earth rotation and its elimination

The position (coordinate value) of the satellite relative to the earth coordinate system is relative to the epoch. The satellite is in a certain position in the coordinate system at a certain signal transmission epoch. When the satellite signal is transmitted to the observation station, due to the earth rotation, the satellite is no longer in the position at the transmission epoch. At this time, in order to eliminate the influence of the earth rotation, the earth rotation correction should be considered to determine the satellite position.

6.   Earth tide effect and its elimination

The gravitation of the sun and moon will lead to the tide phenomenon in the ocean, which will lead to the redistribution of the sea water mass and the additional potential of the ocean tide. The changes of these additional positions lead to periodic deformation of the ground monitoring stations, especially in the offshore areas. The vertical estimation changes can reach several centimeters. The distribution of ocean tide load is related to the distribution of global sea tide height. The rise and fall of sea tide is extremely complex, but its fundamental power source comes from the sun and moon.

    Under the combined action of earth tide and ocean load tide, the maximum vertical displacement of the observation station can reach to 80 cm, which leads to periodic changes in positioning results of GNSS at different epochs. Therefore, in the large-scale high-precision relative positioning or differential positioning work, we must use the earth tide correction error model to correct, in order to obtain high-precision three-dimensional positioning results.

1.5   Satellite Navigation Augmentation System

1.5.1   Satellite Based Augmentation System

Satellite navigation system augmentation system is developed due to the SA (Selective Availability) policy of GPS. In 2000, the United States cancelled the SA policy, and the navigation and positioning accuracy has been improved to a certain extent. With the continuous promotion and deepening of the application of GNSS, the existing satellite navigation system still cannot meet the requirements of some high-end users in terms of positioning accuracy, availability and integrity. Therefore, various satellite navigation augmentation systems are developed.

1.   WAAS

Wide area augmentation system (WAAS) is a satellite based augmentation system for GPS based on the navigation requirements of FAA (Federal Aviation Administration). WAAS includes three parts: one is to provide L-band ranging signal, the other is to provide GPS differential correction data, and the third is to provide integrity information. On this basis, the navigation accuracy, integrity and availability of GPS are improved.

WAAS was officially put into operation in July 2003. It is composed of 38 reference stations (including 9 in North America, which is not in the United States), 3 main control stations, 6 uplink stations, 2 operation control centers and 3 geostationary satellites (not included in GPS). Among them, three geostationary satellites are located at 133°, 107.3° and 98° west longitude, respectively. 25 ground stations are distributed in the United States according to their needs, and are responsible for collecting all data of GPS satellites. Among them, the three main control stations are located in the east and west coasts of the United States, which are responsible for collecting the orbit error of the satellites and the atomic clock error on the satellites, correcting the signal delay caused by atmospheric and ionospheric propagation, and broadcasting the obtained data through the two geostationary satellites.

WAAS does not have the same function as GPS. Its space part includes two geostationary satellites, therefore, its coverage is not global. At present, WAAS provides signals for the United States. The goal of WAAS is to improve the integrity, availability, continuous service and accuracy of GPS SPS (standard positioning signal), which can provide services for civil aviation, vehicles and individual users.

2.   EGNOS

The joint construction of the European geostationary navigation overlay service (EGNOS) system is proposed by the European Space Agency, the European space navigation safety organization and the European Commission. The implementation of EGNOS began in 1998, and its satellite experimental platform was put into use in February 2000.

EGNOS system is the same as WAAS in the United States in principle, covering the whole Europe. EGNOS system can provide high-precision navigation and positioning services for European radio navigation users. The system includes three geostationary satellites and a ground station network. The positioning signals sent by satellites are similar to those of GPS and GLONASS. But integrity information is added to the signals of EGNOS, including the position of each GPS and GLONASS satellite, the accuracy of the atomic clock on the satellite and the ionospheric interference information that may affect the positioning accuracy. The three geostationary satellites are IN-MARSAT AOR-E (15.5° W), ARTEMIS (21.3° W) and IN-MARSAT IOR-W (65.5° E). The ground segment consists of 34 ranging and integrity monitoring stations (RIMS), 4 control centers and 6 navigation ground stations. The system has good stability and high precision. In practical application, the positioning accuracy of the system is better than 1 m, and the reliability reaches to 99%.

3. MSAS

Japanese multi-functional satellite augmentation system (MSAS) is organized and built by Japan Meteorological Agency and Japan Transportation Agency. It is a kind of GPS external enhancement system similar to WAAS of the United States. The difference is that MSAS adopts the multi-function transportation satellite (MTSAT) launched by Japan itself, and its main purpose is to provide full range communication and navigation services for aircraft in Japanese airfield. The MTSAT is equipped with a navigation signal repeater, which transmits the navigation enhancement signal broadcast by the ground reference station. The system can cover Japan, Australia and other regions.

MTSAT is a kind of geostationary satellite, located at 40 and 145° E. MSAS is composed of two satellites. MTSAT 1R satellite was launched on February 26, 2005, and MTSAT 2 satellite was launched on February 18, 2006. MTSAT uses two frequency of Ku band and L-band. The frequency of Ku band is mainly used to broadcast high-speed communication information and meteorological data, and the frequency of L-band is the same as that of GPS L1, which is mainly used for navigation services.

4. GAGAN

Indian GPS aided GEO augmented navigation (GAGAN) system is designed to meet the growing needs of air traffic navigation and enhance aviation navigation capability. GAGAN system will improve security, improve the use of airports and airfield under adverse weather conditions, enhance reliability and reduce flight delays.

The GAGAN system was jointly developed by Airport Authority of India (AAI), Indian Space Research Organization (ISRO) and Raytheon. AAI is responsible for the construction of ground infrastructure, including base station, uplink station and main control center. The construction of GAGAN system mainly includes two stages: technology demonstration system (TDS) and final operational phase (FOP). In the TDS phase, it mainly completes the system index allocation, system integration and on orbit test. The content of the test in this phase is mainly the accuracy index of the system, excluding the integrity information and safety of life (SOL) test. Based on the completion of TDS phase, FOP uses three geostationary satellites to enhance the GPS signal, completes the final integration and put into operation, and can demonstrate the integrity information and SOL service of the system.

GAGAN system consists of space segment and ground segment. The space segment is the GPS dual frequency (L1 and L5) navigation payload of GSAT 4 satellite, and the frequency of C-band and L-band is used as carrier. Among them, C-band is mainly used for measurement and control, and L1 and L5 frequencies of L-band are the same as L1 (1575.42 MHz) and L5 (1176.45 MHz) of GPS, which are compatible and interoperable with GPS. The space signal covers the whole India and can provide GPS information and differential correction information for users. The ground section is composed of eight Indian base stations, one Indian main control center (INMCC), one Indian land uplink station (INLUS) and related navigation software and communication links.

5. SDCM

Russian system of differential correction and monitoring (SDCM) is similar to American WAAS and European EGNOS. It can monitor the integrity of GPS and GLONASS, and provide the analysis results of GLONASS. The system consists of two parts: Ground-based reference station network and two geostationary orbit relay satellites. The horizontal positioning accuracy can reach 1–1.5 m, the vertical positioning accuracy can reach 2–3 m, and the real-time positioning accuracy near the base station (within 200 km) can reach to centimeter level. SDCM plans to build 19 ground reference stations. Two space relay satellites, Ray 5A and 5B, are developed by Leshetnev research and product center in Krasnoyarsk. These two satellites can provide GLONASS correction data and are deployed at 16°W and 95°E, respectively.

SDCM can cover the whole territory of Russian Federation. At present, the Russian government plans to establish monitoring stations outside Russia to improve the integrity, accuracy and reliability of GLONASS.

1.5.2 Ground Based Augmentation System

1. HA-NDGPS

National differential GPS (NDGPS) is a ground augmentation system operated and maintained by the Federal Railway Administration, the Coast Guard and the Federal Highway Administration. It provides more accurate and perfect GPS navigation services for users on the ground and on the water. The modernization work includes the high accuracy NDGPS (HA-NDGPS) which is being developed. This system can be used to enhance the performance and make the accuracy of the whole coverage reach to 10–15 cm. NDGPS is built in strict accordance with international standards, and more than 50 countries in the world have adopted similar standards.

2. LAAS

Local area augmentation system (LAAS) is a navigation augmentation system which can provide high precision GPS positioning service in local area. Its principle is similar to wide area augmentation system (WAAS), except that the GEO satellite in WAAS is replaced by ground reference stations. Through these reference stations, ranging signals and differential correction information are sent to users, like aircraft precise approach.

3. IGS

The predecessor of the international GNSS service (IGS) is the International GPS service organization. The high quality data and products provided by IGS are used in many fields such as geoscience research. IGS is composed of satellite tracking station, data center, analysis and processing center. It can provide high-accuracy GPS data and other data products in real time on the Internet, and can meet the needs of a wide range of scientific research and engineering fields.

4. CORS

Continuous operation reference station system (CORS) is a widely used method for foundation augmentation. Its working principle is to select some reliable GNSS stations from the same batch of GNSS points and have control significance over the whole survey area for continuous tracking observation for a long time. Through the network solution composed of these stations, the "local precise ephemeris" and other correction parameters covering the area and the time period can be obtained for the precise solution of other baseline observations in the survey area.

CORS station meets the needs of long-distance, large-scale centimeter level high-precision real-time positioning. CORS expands the coverage, reduces the operation cost, improves the positioning accuracy and reduces the initialization time of user positioning.

1.5.3 Beidou Ground Based Augmentation System

Beidou ground based augmentation system is a major national information infrastructure, which is used to assist Beidou satellite navigation system to enhance positioning accuracy and improve integrity services. Beidou ground based augmentation system is composed of Beidou base station system, communication network system, national data comprehensive processing system and data backup system, industry data processing system, regional data processing system and location service operation platform, data broadcasting system, Beidou/GNSS augmentation user terminal and other subsystems.

The Beidou ground based augmentation system receives the navigation signals from Beidou navigation satellite through several fixed Beidou reference stations established on the ground according to a certain distance. The signals are transmitted to the data comprehensive processing system through the communication network. After processing, the precise orbit and clock error, ionospheric correction, post-processing data products and other information of Beidou system are obtained, which are transmitted through satellite, digital broadcasting and mobile communication. The Beidou ground based augmentation system meets the real-time positioning and navigation requirements of wide area meter level, decimeter level and regional centimeter level within the service scope of Beidou system, as well as the positioning service requirements of post-processing millimeter level.

1.   Composition of Beidou ground based augmentation system

1)   Beidou reference station network

Beidou reference station network consists of two parts: frame network and regional enhanced density network. The frame network reference stations are evenly distributed on land, coastal islands and reefs of China, which can meet the networking requirements of Beidou ground based augmentation system to provide wide area real-time meter level, decimeter level augmentation services and millimeter level high-precision post-processing services. The base stations of the regional enhanced density network are set up by the provinces, municipalities or autonomous regions as the regional units, and are covered according to their respective area, geographical environment, population distribution, social and economic development. Therefore, it meet the networking requirements of Beidou ground based enhancement system for providing regional real-time centimeter level enhancement services and millimeter level high-precision post-processing services.

2)   Communication network system

Communication network system includes frame network, regional enhanced density network and national data integrated processing system/data backup system, which is used for data transmission, network configuration and monitoring of communication network and related equipment between national data integrated processing system and industrial data processing system, Beidou comprehensive performance monitoring and evaluation system, location service operation platform and data broadcasting system.

3)   National integrated data processing system

The national integrated data processing system of Beidou ground based augmentation system is responsible for receiving the observation data stream of Beidou, GPS and GLONASS satellites in real time from Beidou reference station network, generating Beidou reference station observation data files, wide area augmentation data products, regional augmentation data products, post-processing high-precision data products, etc., and pushing them to industry data processing system, location service operation platform, data processing platform and broadcast system.

4)   Industry data processing system

The industry data processing system includes six industry data processing subsystems of the Ministry of Transport, the Ministry of Natural Resources, the China Earthquake Administration, the China Meteorological Administration and the Chinese Academy of Sciences, and the national Beidou data processing backup system. The six industry data processing subsystems receive the observation data and enhanced data products generated from the Beidou benchmark station of the national integrated data processing system, reprocess the enhanced data products according to the industry characteristics, and form the enhanced data products that support the in-depth application of various industries. The national data processing backup system of Beidou ground based augmentation system provides basic remote data backup service for the observation data of Beidou ground based augmentation system reference station network, so as to ensure that the remote backup system can be restored when the observation data of the national data comprehensive processing system is lost or damaged.

5)   Data broadcast system

The data broadcasting system receives all kinds of enhanced data products generated by the national data comprehensive processing system, processes and encapsulates the broadcasting requirements of all kinds of data products, and then transmits the processed and encapsulated enhanced data products to the user terminal or receiver for use by all kinds of broadcasting means. Data broadcasting system uses satellite broadcasting, digital broadcasting and mobile communication to broadcast enhanced data products.

6)   Beidou/GNSS enhanced user terminal

Beidou/GNSS enhanced user terminal (receiver) is used to receive signals from Beidou satellites and enhanced data product signals from data broadcasting system, so as to realize high-precision positioning and navigation functions.

2.   Service products of Beidou ground based augmentation system

Beidou ground based augmentation system now provides wide area augmentation service, regional augmentation service and post-processing high-precision service, respectively corresponding to wide area augmentation data products, regional augmentation data products and post-processing high-precision data products. Wide area augmentation data products and regional augmentation data products provide services through mobile communication. The post-processing high-precision data products can be downloaded as files.

Wide area enhanced data products include Beidou/GPS precise orbit correction, clock error correction, ionosphere correction, etc.

Regional enhanced data products include Beidou/GPS/GLONASS regional comprehensive error correction.

Post processing high-precision data products include Beidou/GPS post-processing precision orbit, precision clock error, EOP, ionospheric products, etc.

3.   Service performance index of Beidou ground based augmentation system

Table 1.9 Beidou wide area positioning accuracy index

Product classification	Positioning accuracy (95%)	Constraint condition
Wide area enhanced data products	Single frequency pseudo range positioning: Horizontal $\leq$2 m Vertical $\leq$4 m	The number of Beidou effective satellites is more than 4 PDOP < 4
	Single frequency carrier phase precise single point positioning: Horizontal $\leq$1.2 m Vertical $\leq$2 m	The number of Beidou effective satellites is more than 4 PDOP <4
	Dual frequency carrier phase precise point positioning: Horizontal $\leq$0.5 m Vertical $\leq$1 m	The number of Beidou effective satellites is more than 4 PDOP <4 The initialization time is 30–60 min

(1)   Wide area enhanced precision service covers land and territorial waters of China within the broadcast range.

(2)   The service scope of regional enhanced precision refers to the site distribution of regional enhanced density, and the service scope published by the regional service system.

(3)   The high-precision post-processing service covers land and territorial waters of China within the broadcast range.

The positioning accuracy refers to the statistical value of the difference between the given position and the real position of the user, including horizontal positioning accuracy and vertical positioning accuracy. The positioning accuracy index of Beidou ground based augmentation system is shown in Tables 1.9, 1.10, 1.11 and 1.12. If the continuous observation time is not specified, the default is the positioning accuracy index after 24 h of continuous observation.

1.6   Beidou Short Message

The biggest characteristic of Beidou system is its short message communication service. At present, Beidou system is the only navigation system that can carry out short message communication. From the beginning of its birth, Beidou pioneered the integration of positioning, navigation, time service and position report short message functions, which is different from other GNSS. Beidou short message has played an important role in life rescue and special industries. For example, during the Wenchuan earthquake rescue in 2008, the location information is reported to the rescue center through Beidou short message.

Table 1.10 Precision index of Beidou/GPS integrated wide area positioning

Product classification	Positioning accuracy (95%)	Constraint condition
Wide area enhanced data products	Single frequency pseudo range positioning: Horizontal ≤2 m Vertical ≤3 M	The number of Beidou effective satellites is more than 4 The number of effective GPS satellites is more than 4 PDOP <4
	Single frequency carrier phase precise single point positioning: Horizontal ≤1.2 m Vertical ≤2 m	The number of Beidou effective satellites is more than 4 The number of effective GPS satellites is more than 4 PDOP <4
	Dual frequency carrier phase precise point positioning: Horizontal ≤0.5 m Vertical ≤1 m	The number of Beidou effective satellites is more than 4 The number of effective GPS satellites is more than 4 PDOP <4 The initialization time is 30–60 min

Table 1.11 Regional positioning accuracy index

Product classification	Positioning accurac y (RMS)	Constraint condition
Regional enhanced data products	Horizontal ≤ 5 cm Vertical ≤10 cm	The number of Beidou effective satellites is more than 4 or The number of GPS effective satellites is more than 4 or The number of effective satellites of GLONASS is more than 4 PDOP <4 Initialization time ≤60 s

Table 1.12 Post processing positioning accuracy index

Product classification	Positioning accuracy (RMS)	Constraint condition
Post processing high precision data products	Horizontal ≤ 5 mm ± 1 ppm × D Vertical ≤ 10 mm ± 2 ppm × D	The number of Beidou effective satellites is more than 4 or The number of effective GPS satellites is more than 4 PDOP <4 Continuous observation for more than 2 h

1.6.1   Characteristics of Beidou Short Message Communication

1.   Communication link

Beidou short message communication links users with each other through satellites. Users establish communication links with other users through Beidou satellite. It is similar to the link layer of Internet. The link layer defined in satellite TCP/IP transmission technology is not only the communication link of the whole system, but also a higher level on this basis. The link control function is not realized in the actual link. There are problems of data loss and propagation delay, as well as information asymmetry.

2.   The limitation of communication frequency and volume

For short message service of Beidou system, the early user communication capacity is 36 Chinese characters at one time. The current user communication capacity is 120 Chinese characters at one time. The frequency of providing service varies according to the classification. The slowest is 10 min, and the fastest is 1 s.

3.   Type of data format

There are two kinds of data formats used in short message communication of Beidou system. One is ASCII code for Chinese character communication, the other is BCD code.

4.   Interference and restriction factors in communication process

Beidou short message is easily affected by weather and other environmental factors. Its data volume and frequency restrict its flexibility. The bit error rate of data transmission is high. Therefore, it is more suitable for special industries such as emergency rescue.

1.6.2   Beidou Short Message Communication Mode

1.   Communication between users

The short message sent by the Beidou client user directly arrives at the Beidou client user through the satellite channel. The client user can be divided into the main card and the sub card. When the sub card sends the short message, it will send a short message to the main card commander user at the same time. The main card commander user can broadcast the short message to all the sub cards, which is similar to the function of short message group sending. This function can be applied to the weather broadcast and emergency communication for ships. Because the communication frequency of Beidou short message is limited to 1 time/minute, the general user will control the short message by the way of queue and send it one by one in order. But there is no time limit for the commander user or the client user to receive the short message.

2.   Communication between client user and ordinary mobile phone

The client user of Beidou needs to be forwarded by the communication service of the commander user before sending short messages to ordinary mobile phones. First of all, the Beidou client user sends the short message to the commander user. The communication service of the commander user receives the short message through the serial port to judge the firs 11 numbers of the short message whether they are the mobile phone number. The Beidou commander user will push its short message to the SMS gateway through the network by identifying the mobile phone number. Then send it to the target mobile phone by the SMS gateway, thus, the short message communication from the Beidou client user to the ordinary mobile phone without signal and network coverage is done. On the contrary, ordinary mobile phones can also send short messages to Beidou client users. After receiving the short message from the mobile phone, the communication service of the commander user judges the sending target by identifying the first six bits of the short message content. By calling its interface, the commander user sends the short message to the client user to realize the function of sending the short message from the ordinary mobile phone to the Beidou client user.

3.   Emergency rescue communication

Beidou short message emergency channel is set, which can continuously send distress signals according to certain time interval without time limitation. In general, the short message for emergency rescue provides hardware or software button, which can be provided to users in the simplest and fastest way for emergency usage.

1.6.3 Application of Beidou Short Message Service

The short message service of Beidou system can send out the location information of users, so that the third party can acquire the use's situation. It has very important military and civil value, and has broad application prospects. The service fields involve emergency communication, location monitoring, data transmission, etc., including fishing vessel location monitoring, wildlife location tracking, outdoor and maritime emergency rescue, meteorological monitoring, etc. Each application can make full use of Beidou short message to achieve satellite communication in the blind area without mobile phone signal, with low communication cost. Through the customization of terminal products and systems, various solutions are formed to truly apply the short message technology to specific scenarios to solve the pain points in practical application.

At present, the combination of Beidou short message technology and mobile Internet technology has opened up more application scenarios, and truly achieved the civil communication function of Beidou short message. Smart phone apps can solve the communication problem in the area without public network signal. After downloading and installing the apps, users can use their smart phone to connect with the Beidou terminal with RD module through Bluetooth, which can solve the problem of seafarers and fishermen's communication at sea. In addition, they can also provide mobile signal interruption to emergency rescue service units. In case of earthquake or disaster, it can provide text messages for emergency rescue, or provide people who like to go hiking in remote areas with services such as querying the nearest parking space, restaurant, hotel, etc., while providing rescue services in case of distress without information coverage. When the search and rescue is carried out in the desert, remote mountainous area, sea and other sparsely populated areas without signal coverage, the Beidou terminal can also timely report the location and disaster situation of the victims. The rescue efficiency can be effectively improved.

Chapter 2
Internet of Things

Internet of things (IOT) is a kind of network that collects real-time information through various information sensing devices, connects any real object in the physical world with the Internet according to the pre-defined protocol. IOT carries out information transmission and interaction, to achieve the purpose of intelligent identification, accurate positioning, tracking and control of information.

The IOT has three basic characteristics: interconnection, deep perception and intelligent service. The basic feature of the IOT is interconnection. Firstly, the IOT should be able to meet the communication requirements of a variety of heterogeneous networks, heterogeneous terminals and sensor devices, to prepare for the next step of sensing and interaction of a large amount of data. Deep sensing is the key of the IOT, which provides the processed sensing data through multi-level fusion and collaborative processing of physical sensing information. According to the information to the application layer, the IOT should be able to achieve the function of intelligent services. According to the results of real-time perception and information fusion of the physical environment, the IOT makes its own decisions to dynamically adapt to the real-time changing physical world.

2.1 IOT Architecture

The IOT should have three significant functions: comprehensive perception, reliable transmission and intelligent processing. Comprehensive perception refers to the use of radio frequency identification (RFID), sensors, QR code and other sensor devices to comprehensively collect the object information of the physical world. Reliable transmission is the real-time and accurate transmission of the perceived information through a variety of communication networks and the Internet. Intelligent processing is the use of cloud computing, fuzzy identification and other intelligent computing technologies in the analysis and processing of a large amount of data to achieve intelligent control.

© Publishing House of Electronics Industry 2022
B. Wang et al., *Internet of Things and BDS Application*,
https://doi.org/10.1007/978-981-16-9194-2_2

2.1.1 Technology Architecture of IOT

1.   Three layer architecture of IOT

The architecture of IOT can be divided into perception layer, network layer and application layer, as shown in Fig. 2.1.

(1)   The perception layer is the main part of object identification and data collection. It is at the bottom of the three layer architecture of the IOT. It is equivalent to the nerve endings of human perception organs and is used to perceive external things. The sensing layer is mainly composed of sensors and sensor gateways, which collect data through information sensing devices, such as sensors, QR code, RFID and so on, and identify the monitored objects. The key technologies of sensing layer include sensor technology, recognition technology and positioning technology.

In the IOT, the main information collection device is sensor, which is the basis of perception, service and application of the IOT. The sensor converts the information collected from the physical information, chemical information and others into a certain form of electrical signal by using the relevant mechanism. The sensors process the signal through the corresponding signal processing device, and produces the corresponding action. Sensing technology is to convert analog signal into digital signal and quantify information. In the IOT, sensor nodes are composed of multiple modules, including sensing, information processing, sending and receiving network information, providing energy and so on. Compared with the traditional sensors, the sensors in the IOT have additional functions such as collaboration, computing and communication. A large number of sensor nodes are arranged in the detection range, and the wireless network system is called wireless sensor network. Wireless sensor network can sense, collect and process the monitoring information of the network area through the cooperation of each sensor node, and transmit it to the user.

Automatic identification technology unifies the physical world and the information world, which is the main feature of the IOT different from other networks.

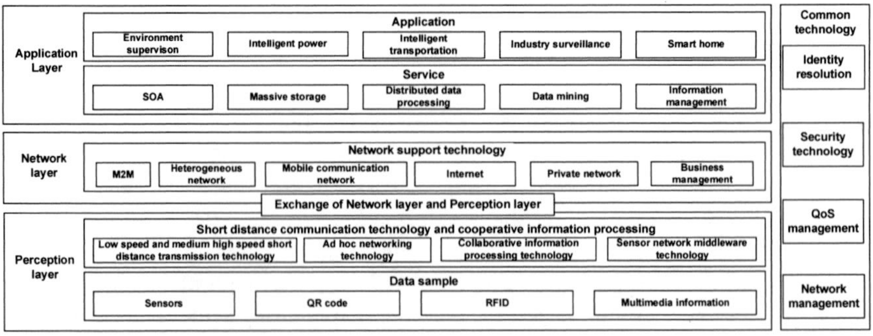

Fig. 2.1 Three layer architecture of IOT

Automatic identification technology is a kind of technology that uses specific identification equipment to identify the proximity activity of object middleware, so that
it can independently extract the data information related to the identified object,
and provide the information to the upper layer equipment for further processing.
Through the automatic identification technology, the IOT can automatically collect
data, identify the information, input it into the computer, and assist people to complete
the real-time analysis of a large number of data.

Positioning technology is a kind of technology that uses a certain algorithm to
measure the location of people, objects and events in the specified coordinate system.
It is one of the main research directions of the development and application of the
IOT. The main positioning technology used in the IOT mainly includes GNSS, WiFi,
ZigBee, RFID and so on.

(2) The network layer of the IOT is the key part. The communication network
    analyzes and processes the sensed information transmitted from the sensing
    layer, transmits it to other networks, and feeds back the control commands
    to the sensing layer. Because the network layer can limit the sensing layer
    data to a specific area for transmission, it is also called the transmission
    layer. The network layer is equivalent to the nerve center and brain of
    the IOT. It is composed of various private networks, Internet, communica-
    tion networks, etc., which makes the data between the end-to-end efficient
    and barrier free transmission. Its transmission reliability is high, and the
    transmission process is relatively safe. It can carry out a wider range of
    data interconnection, and has the functions of addressing, routing, network
    connection, data holding/interruption, etc. The network layer needs to use
    the embedded system to physicalized its network. The embedded system can
    control its physical devices through computing technology. The key technolo-
    gies in the network layer include long-distance wired/wireless communication
    technology, network technology, etc.

(3) The application layer processes and transmits the information from the former
    two layers, controls and makes decisions, and processes the functions of infor-
    mation storage and data mining. It can achieve the purpose of intelligent
    management, application and service. The application layer of IOT is composed
    of applications in different industries, such as medical, agricultural, etc. The
    application layer provides users with interfaces according to their specific needs
    to complete the information interaction and sharing among different industries,
    applications and systems, so as to achieve the real intelligent application of
    IOT. After the information data is transmitted by the network layer, the appli-
    cation layer analyzes and processes the relevant data to form the information
    needed by users, and can process part of the information to provide rich specific
    services. According to different functions, the application layer can be divided
    into service support sublayer and application sublayer. The service support
    sublayer constructs a dynamic resource database that can meet the require-
    ments of real-time update by using the information data transmitted from the
    lower layer. The main role of the application sublayer is to modify the data

resources, network and perception layer technologies according to the different needs of the industry, and make different solutions.

The IOT has a wide range of applications, such as the monitoring of environmental pollution, intelligent retrieval, intelligent life, device control, payment services, etc.

The IOT is a large-scale information system, which needs to deal with a huge amount of data. An important problem to be solved in the application layer of the IOT is how to properly process the massive data and extract the effective information. The important technologies in the application layer of the IOT include artificial intelligence technology, cloud computing technology, M2M platform technology, etc.

1)   Artificial intelligence technology

Artificial intelligence technology can improve automation and intelligence of machine, thus enhancing the operating environment and reducing the workload. Artificial intelligence technology can also improve the reliability of equipment, reduce the cost of maintenance and operation, and carry out intelligent fault diagnosis.

2)   Cloud computing technology

When the IOT is dealing with massive data, cloud computing can decompose the computing processing program in the network into multiple subprograms, and then process and analyze the data information through the system composed of multiple servers, and send the results back to the client. Cloud computing technology integrates different computing methods, information virtualization, load balancing and other functions of traditional computer technology and traditional network technology. Cloud computing has strong data processing ability, high reliability of computing results, strong storage capacity, high cost performance in computing technology, and is very suitable for IOT applications. For a variety of IOT applications, cloud computing technology builds a unified platform for service delivery, improves the computing method and resource storage method, and provides a unified data storage format and data processing method. Cloud computing can greatly simplify the application delivery process, reduce the delivery cost and improve the processing efficiency. Figure 2.2 shows the cloud computing model.

3)   M2M platform technology

M2M (machine to machine) is the communication between machines. M2M focuses on the research of wireless communication between machines, mainly including machine to machine, machine to mobile phone (user remote monitoring), mobile phone to machine (user remote control). The communication method of M2M is intelligent and interactive. The machine can initiate communication according to the established program, instead of waiting for communication passively, and can make intelligent decisions according to the obtained data, and send decision instructions to the corresponding equipment. Figure 2.3 shows the structure of M2M system.

M2M products are usually composed of three parts: wireless terminal, transmission channel and application center. The application terminal of specific industry

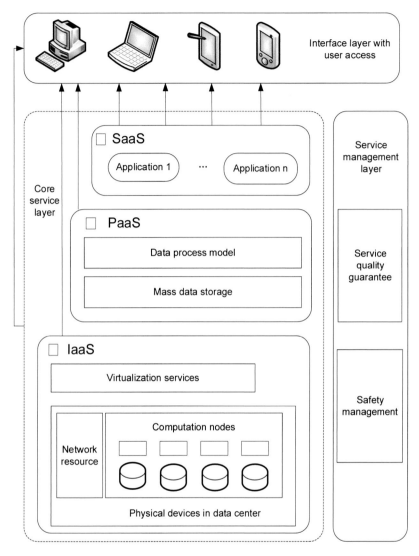

Fig. 2.2 Cloud computing model

is wireless terminal, not the terminal equipment mentioned in the general Internet, such as mobile phone, laptop, etc. The data on the terminal is sent to the application center through the transmission channel to complete the data summary. The application center can control the distributed wireless terminal according to the summarized data.

In addition to the above forms of IOT architecture representation, there are also "sea-network-cloud" or "end-channel-cloud" architecture representation methods. "End" refers to the terminal equipment, which is all IOT sensing terminals and

Fig. 2.3 M2M system structure

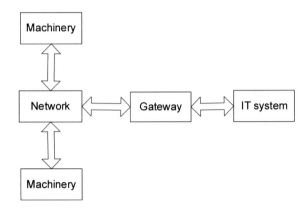

user terminals. "Channel" refers to the transmission channel, which is equivalent to the network transmission layer. "Cloud" refers to the cloud, which includes the application in the scope of cloud services, which is equivalent to the application layer. Different architecture representation methods are only from different perspectives, but the essence of the IOT is the same.

2.   Four layer architecture of IOT

The four layer architecture is mainly composed of perception layer, network layer (supply layer), support layer and application layer. The four tier architecture of the IOT is shown in Fig. 2.4.

1)   Perception layer

The sensing terminal sublayer is equivalent to the neural terminals of the IOT. Its main task is to conduct reliable sensing, that is, to collect and process the surrounding objective physical environment parameters. The sensing convergence sublayer contains all kinds of wired or wireless field networks. Its main task is to carry out signal transmission and aggregation. Its core device is the gateway of IOT, which has the function of field network management and is responsible for forwarding the information of field network and various WAN.

2)   Network layer (transmission layer)

Network adaptation sublayer judged the target network of data and generated the corresponding protocol data unit. The transmission bearing sublayer includes various IP private network, metropolitan area network, CMNET, etc., which is the main bearing layer of data information transmission. The core network sublayer divides the network into circuit domain, packet domain, CM-IMS domain, etc. according to the different forms of transmission information.

3)   Supporting layer

Fig. 2.4 Four layer
architecture of IOT

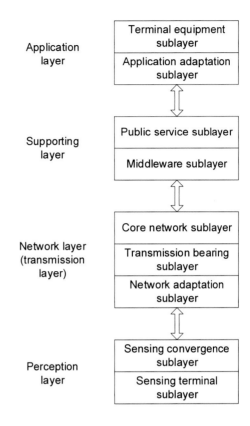

The middleware sublayer is composed of various service middleware, which can extract the characteristics of the basic performance set of various specific applications, and provide the service invocation function to the upper public service layer. The public service sublayer contains matching programs of different industries, provides industry solution planning, calls the common service interface provided by the middleware sublayer to other programs. The upper application layer uses data results of public service sublayer to make the decision and control.

4)   Application layer

The application adaptation sublayer unifies the coding format of all application layer information data and provides a unified interface. The terminal equipment sublayer provides virtual table for data query, and the specific application of each industry finally completes the decision in this layer. The decision maker issues the instruction code according to the demand.

3.   Five layer architecture of IOT

As an open architecture using open protocols, the IOT can support a variety of Internet based applications. In order to promote the information interaction and integration between the Internet and real world, the IOT should also have scalability, network security, semantic representation and other functions. Therefore, scholars have proposed a five layer architecture system for qualitative analysis of the IOT, rather than specific definition of its protocol. The physical model includes five layers: perception control layer, network interconnection layer, resource management layer, information processing layer and application layer.

(1) The sensing control layer is composed of RFID reader, intelligent sensor node and access gateway, etc. The IOT perceives the external information through each sensor node in the sensing control layer. The sensor nodes can compose a network by themselves and access the sensing data to the gateway. Finally, the gateway transmits the data to the Internet for further data processing and analysis.

(2) The network interconnection layer uses various access devices to interconnect different networks, such as Internet and communication network, to achieve the functions of data format, address translation, transmission and communication.

(3) The resource management layer initializes the resources, monitors the operation of resources, coordinates the work among multiple resources, and achieve the interaction of cross domain resources.

(4) The information processing layer can achieve the analysis and reasoning of perception data, and decision-making. The information processing layer can also complete the query, storage, analysis, mining and other operations of data.

(5) The application layer can provide users with a variety of services.

In addition to the foundation of the above five layers of architecture, there are various mechanisms in the IOT that can serve as the support of the basic architecture and serve specific applications, such as security mechanism, fault tolerance mechanism and service mechanism.

2.1.2 IOT Platform Architecture

The IOT platform is based on Internet and communication technology. On this platform, users can access the IOT through their own original device and technology, without using specific hardware modules. The platform architecture of IOT consists of four core modules, network and intelligent devices. The four core modules include device management, user management, data transmission management and data management. These four modules are the basis of the IOT platform, and other functional modules are based on the extension of these four modules. Figure 2.5 shows the architecture of the IOT platform.

1. Device management

Device management can be divided into two parts, one is the type management of device. This module mainly defines the type of device, which is generally completed

Fig. 2.5 Architecture of
IOT platform

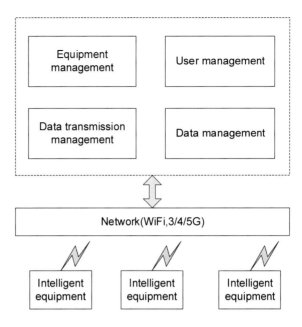

by the device manufacturer, that is, the device manufacturer defines the data analysis method, data storage method, equipment specification and other information. The user of the device can only browse the related information of the device, but cannot define the data. The other part is the information management of the device, which can define the information about the device. The device user has full control over the device activated after purchase, and can control which data of the device can be viewed by the manufacturer and which data can be viewed by the user.

2.   User management

User management can be divided into organization management, personnel management, user group management and authority management.

(1)   Organization management: on the IOT platform, all devices, users and data are based on organization management, while organizations can be device manufacturers, device users or families.
(2)   Personnel management: users are composed of personnel in an organization. There are administrators in each organization. Administrators can add different users to the organizations they serve and assign different permissions to each user. The same user may exist in multiple different organizations.
(3)   User group management: enable users in the same user group to have the same permissions.
(4)   Authority management: authority management is mainly the specific division of authority.

3.   Data transmission management

Data transmission management is a data transmission protocol defined for a class of devices. Its basic format is

Device serial number@Command code@Data

where the device serial number is the number specified by the manufacturer for each device, and there is no fixed format, which depends on the coding format of the manufacturer; the command code can reflect the function of this data, such as uploading data, or the command sent to the device by the server, which is generally encoded by two digits 00–99; the data is the data part contained in this message. Each protocol can define different parsing methods. Each device type can define multiple commands. Each command has a different parsing method. The organization administrator can define the parsing method according to the requirements. After the server receives the data, it will automatically parse the data fields according to the pre-defined parsing method. The data fields are sent and received according to the HEX method. According to the data format defined by the IOT platform, developers develop their own device parsing code.

After the completion of data analysis, the data should be stored. The IOT storage should apply the distributed architecture, so that each device can be specified with a different storage location. In Diego IOT, MySQL database is used for data storage. Different devices can be stored in different MySQL databases, and the life cycle of each data is defined. At the end of the life cycle, the data will automatically be delete by the system.

4. Data management

There is a large amount of data in the IOT. We can use the open source big data platform to achieve data visualization analysis and get valuable data. Permission management is an important part of IOT data management. Only the device owner can define the data browsing permission. Users can also export the data to the local and do the analysis.

2.1.3 Network Communication Architecture

At present, the communication between the IOT platform and devices in the cloud is based on TCP/IP protocol. On this basis, the communication between devices and cloud platform can also use WiFi, 4G, 5G and other methods, while the communication between devices can use WiFi, Bluetooth, ZigBee and other methods.

1. Communication based on 4G/5G mobile signal

The communication architecture of 4G/5G is the simplest one. The following should be considered.

Fig. 2.6 Communication process based on mobile 3/4/5G

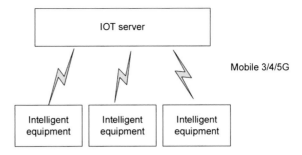

(1) Each device needs a SIM card. You can go to the mobile operator to apply for an IOT SIM card.

(2) Attention should be paid to data traffic. This architecture is based on 4G/5G communication. Therefore, it will completely consume the mobile data traffic. If there is video data, it will produce relatively large traffic charges.

(3) Communication quality needs to be considered. This architecture relies on the network coverage of mobile operator. Data cannot be sent and received in some weak or non-signal environments.

Figure 2.6 shows the communication process based on mobile 3G/4/5G.

2. Communication based on WiFi LAN

The communication architecture based on WiFi LAN is suitable for all intelligent devices of IOT working in a local environment. The devices are connected to the router through WiFi or wired LAN, and then connected to the IOT server by the router. The intelligent device in the LAN has no independent IP of the public network, only one IP in the LAN. Therefore, the intelligent device can directly send data packets to the IOT server, the Internet of things server cannot directly send data packets to the device. Due to the high power consumption of WiFi, it is necessary to consider the power supply of intelligent devices connected through WiFi.

Interference is also one of the issues that need to be considered in this architecture. If there are strong interference sources in the environment, such as electromagnetic interference, it is necessary to use routers with strong anti-interference ability. Figure 2.7 shows the communication process based on WiFi or wired LAN.

3. Communication based on Bluetooth

In this architecture, the intelligent devices are connected to the gateway through Bluetooth, and then connected to the IOT server by the Bluetooth gateway. Bluetooth is a point-to-point communication mode. The following problems should be considered.

(1) Bluetooth gateway capacity, which is how many Bluetooth devices a Bluetooth gateway can access.

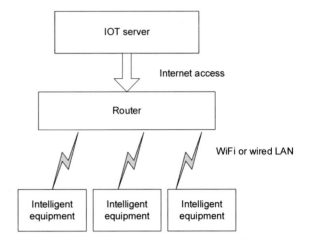

Fig. 2.7 Communication process based on WiFi or wired LAN

(2) Bluetooth pairing. Bluetooth devices need to pair before communication. If it cannot automatically pair, then in the case of more intelligent devices in the IOT, the communication method based on Bluetooth is not appropriate.

There is also a case for the IOT devices that do not need to be online all the time. Only in some special cases, the IOT devices need to be connected to the server. In this case, the devices can be connected to the IOT through Bluetooth. Bluetooth bracelet is a typical application mode of this architecture. Figure 2.8 shows the diagram of the communication process based on Bluetooth.

4. Communication based on ZigBee

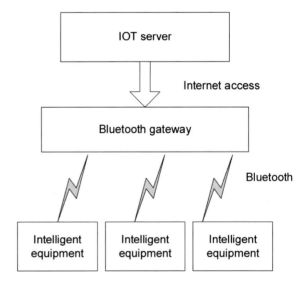

Fig. 2.8 Communication process based on Bluetooth

Fig. 2.9 Communication process based on ZigBee

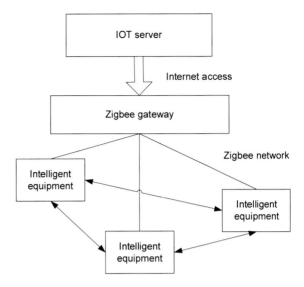

ZigBee is a kind of networking method for sensors, which has very low power consumption ability. ZigBee depends on ZigBee gateway to access network, and the gateway itself is a ZigBee device. ZigBee is the ad hoc network. We should pay attention to the amount of data during usage. ZigBee is an ultra-low power wireless communication technology. The communication ability and power loss of the device change inversely, thus, the communication ability of ZigBee is weak. It is suitable for data acquisition of sensors and other applications with small amount of data, while it is not suitable for large amount of data. Figure 2.9 shows the communication process based on ZigBee.

2.1.4 Gateway Architecture of IOT

In the IOT, the gateway of IOT is very important. Its main functions are network isolation, protocol conversion, adaptation and data transmission inside and outside the network.

After the IOT devices access to the network, IOT communication protocol is needed for communication between devices and between devices and cloud. Only the devices that follow the communication protocol can communicate with each other, complete data interaction, and achieve the function of IOT. The common communication protocols of IOT are MQTT, COAP and so on. These communication protocols are based on the message model. Communication is achieved between devices and between devices and cloud by exchanging messages with data. Figure 2.10 shows a typical gateway architecture of the IOT.

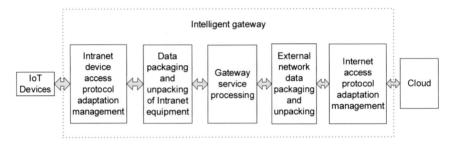

Fig. 2.10 Typical IOT gateway architecture

Fig. 2.11 Terminal system
framework without RTOS

2.1.5   Software System Architecture of IOT Terminal Device

The common software system framework of IOT terminal device is mainly divided
into two types: with RTOS (Real-time operating system) and without RTOS. RTOS
is a real-time multi task operating system. The tasks handled by terminal system
without RTOS are usually single. The terminal system with RTOS can handle multiple
tasks in parallel. Each task is responsible for one transaction, and the efficiency of
system response can be improved by parallelization. RTOS real-time operation kernel
generally includes important components such as task scheduling, synchronization
and communication between tasks, memory allocation, interrupt management, time
management and device driver. Figure 2.11 shows the framework of device terminal
system without RTOS.

Figure 2.12 shows the framework of device terminal system with RTOS.

2.1.6   System Architecture of IOT Cloud Platform

Figure 2.13 shows the system architecture of the IOT cloud platform.

The system architecture of IOT cloud platform mainly includes four components:
device access, device management, rule engine, security authentication and authority
management.

Fig. 2.12 Terminal system framework with RTOS

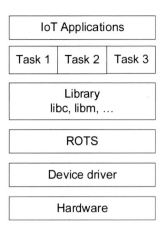

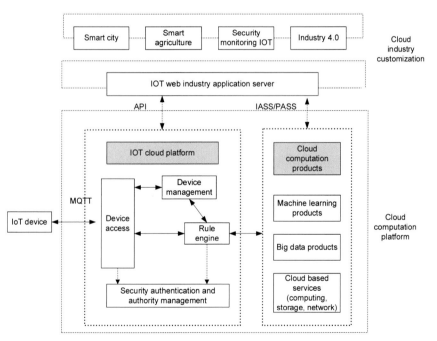

Fig. 2.13 System architecture of IOT cloud platform

1. Device access

Device access component includes a variety of device access protocols, such as MQTT protocol, which can perform concurrent connection management and maintain long-term connection management of billions of devices. Some cloud computing companies also simplify their MQTT protocol and turn it into a unique access protocol. At present, most of the open MQTT proxy servers are stand-alone versions, with a maximum of more than 100,000 devices connected in parallel. Therefore, if you want to manage the connection of billions of devices, you need to use the load balancing and distributed architecture, and also need to arrange the distributed MQTT proxy server in the cloud platform.

2.   Device management

The device management mode is generally tree structure, including device creation management and device status management. Device management takes IOT products as root nodes, then device groups, and then specific devices. Device management mainly includes product registration and management, device addition, deletion, modification and query management, device information release, OTA device upgrade management, etc.

3.   Rule engine

Cloud platform of IOT is built on the basis of cloud computing platform. Cloud platform of IOT and cloud computing platform jointly provide services for IOT service. The rule engine first filters the data of the IOT platform, and then forwards it to other cloud computing products, such as forwarding the data uploaded by the device to the Table Store database. Rule engine generally uses SQL like language. Users can write SQL language to filter and process data, and send data to other cloud computing products or other cloud computing service terminals.

4.   Security authentication and authority management

IOT devices need to carry the only certificate issued by the IOT cloud platform before they can access the cloud platform. Therefore, each device accessing the cloud platform needs to store a certificate locally in the form of a key composed of multiple strings. Each time the device establishes a connection with the cloud, it needs to carry a certificate so that the cloud security components can pass the verification. The minimum authorization of cloud platform is generally device level authorization. Certificate is generally divided into product level certificate and device level certificate. Product level certificate can operate all the devices under the product with the largest operation authority. Device level certificate has less authority, and can only operate its subordinate device, while cannot operate other devices.

2.2   Information Transmission Protocol of IOT

2.2.1   Short Range Wireless Communication Technology of IOT

The core technology of IOT is C^3SD, which includes control system, computing system, communication system, sensing system and data. Communication system and data correspond to the network layer of IOT, while control system and computing system correspond to the application layer of IOT. IOT data can be transmitted safely and reliably through wired transmission and wireless transmission. In the IOT, short-range wireless transmission technology is an important technology.

Wireless communication is a special communication method, which mainly uses the characteristics of electromagnetic wave signal transmission in space to achieve information exchange. Wireless communication mainly includes microwave communication and satellite communication. The distance between the objects to be communicated in the short-range wireless communication is short, while the communication frequency is high. Therefore, the communication equipment can use the non-contact point-to-point data transmission mode. Generally speaking, as long as the data is transmitted by radio and the transmission distance is short, it can be called short-range wireless communication.

The wireless transmission power of the short-range wireless communication technology ranges from 1 μW to 100 mW. The omnidirectional antenna and circuit board antenna are used, and the communication distance ranges from several centimeters to several hundred meters. Short range wireless communication technology can use radio spectrum resources without applying for frequency resource license. It is also applicable in the case of lack of frequency resources. The short-range wireless communication between devices is powered by battery and can be networked with no center.

The short-range wireless communication technologies applied in the IOT include WiFi, Bluetooth, IrDA, ZigBee, NFC, UWB, DECT, etc.

1. WiFi (Wireless Fidelity)

Devices can access wireless LAN by WiFi with RF band 2.4G UHF or 5G SHF ISM. WLAN is generally protected by a password, and the connection to the network needs to pass the verification. The network can also be open to the outside world, allowing any device to access the WLAN within the network coverage. The purpose of WiFi is to improve the communication performance between wireless network devices based on IEEE 802.11 standard. LAN using IEEE 802.11 series protocol is called WiFi.

In WiFi technology, the laying range of radio wave is very large. Its radius can reach about 100 m, which is much larger than that of Bluetooth. The bandwidth of 802.11b wireless network can be adjusted automatically to adapt to the new situation, with the maximum bandwidth is 11 Mb/s. When the signal becomes weak or disturbed, the bandwidth can be automatically changed to 5.5, 2 and 1 Mb/s. In this way, the stability and reliability of wireless network can be improved. The standard protocols of WiFi technology are 802.11a protocol and 802.11g protocol. It works in 2.4 GHz band and the fastest data transmission rate is 54 Mb/s. WiFi provides a technical basis for intelligent devices to access WLAN.

Wireless network is different from wired network. In the range of wireless network, its signal will weaken with the increase of transmission distance, and is easy to be disturbed. The wireless signal is easy to be blocked by objects, resulting in varying degrees of distortion. Radio signals are also vulnerable to the same frequency of radio interference, lightning weather and other environmental effects.

Because WiFi network can directly use the frequency of wireless network without completing the application, the network is easy to be saturated, and vulnerable to illegal intrusion, thus, the security of the network is not high. In order to improve the security of the network, 802.11 protocol proposes an encryption algorithm, which can encrypt the wireless transmission data between the network access point and the host device, namely WEP. This algorithm can effectively prevent the network from being attacked and invaded. Because WiFi network is a kind of wireless network, it has no similar physical structure as wired network for protection. Usually, before accessing the wired network, you need to use the network cable to access. When accessing the wireless network, if the network does not take effective protection, as long as it is within the signal coverage, you can access the wireless network only through the wireless network card, and the security is poor.

WiFi network does not have high requirements for physical lines. There is no need to lay a large number of cables. WiFi technology combined with web service technology can reduce the cost of infrastructure network construction and expand the application scope of WiFi network.

2.   Bluetooth

Bluetooth is a kind of wireless communication technology for short-range information transmission. It uses 2.4 GHz ISM band UHF radio wave to complete short-range data transmission between fixed devices, mobile devices and building personal area network. Bluetooth applies FHSS technology to send data packets through the designated Bluetooth channel. Bluetooth has the master–slave structure protocol, which can set temporary peer-to-peer connection. Bluetooth devices in the network can be divided into master and slave. Through Bluetooth, multiple devices can be connected at the same time to solve the problem of data synchronization.

Bluetooth technology combines circuit switching technology and packet switching technology. Data can be transmitted through asynchronous data channel, three-way voice channel and asynchronous data and synchronous voice parallel transmission channel. PCM or CVSD method is used to modulate the voice signal code. The data transmission rate of single voice channel is 64 kb/s. When asymmetric channel is used to transmit data, the data can be divided into forward transmission and reverse transmission according to its transmission direction. The fastest forward transmission rate is 721 kb/s. The fastest reverse transmission rate is 57.6 kb/s. The highest transmission rate in symmetric channel is 342.6 kb/s.

The power loss of Bluetooth device is relatively small. There are four working modes of Bluetooth device in communication connection state, namely activation, breathing, holding and sleeping. The Bluetooth device works normally when it is in active mode, while the other three modes are low power consumption mode for energy saving.

The data transmission rate of Bluetooth is 1 Mb/s, and the transmission distance is about 10 m. ISM is an open band. Bluetooth and other devices with the same band share the ISM band signal, resulting in mutual influence between the signals.

Bluetooth technology is widely used in a variety of data and voice devices in LAN, such as smart home, which is embedded in traditional household appliances, electronic lock, fax machine, digital camera, mobile phone and headset.

3.   IrDA (Infrared Data Association)

IrDA data protocol consists of three basic layer protocols, namely physical layer, link access layer (IrLAN) and link management layer (IrLMP). SIR (IrDA1.0) standard is a serial half duplex system, whose highest data transmission rate is only 115.2 kb/s. In order to improve the data transmission rate, FIR (IrDA1.1) protocol is proposed. The maximum data transmission rate can reach to 4 Mb/s. VFIR technology belongs to IrDA1.1 standard. Its maximum communication rate can reach to 16 Mb/s. IrDA stack supports a variety of protocols to meet the application requirements of each layer, such as infrared physical layer connection specification, infrared connection access protocol, infrared connection management protocol, etc.

IrDA completes the connection through four stages. The first stage is that IrDA searches for intelligent devices in the network coverage, obtains the device address and completes the resolution. The second stage is to decide which devices to establish the connection and send the connection request according to the application requirements. The third stage is to exchange information in the master–slave mode, and the master device controls the operation of the slave device. The fourth stage is to disconnect the master and slave devices after data transmission.

IrDA module is small in size, low in power loss, easy to connect, simple in structure and easy to use, which can meet the needs of mobile communication. IrDA can directly use the frequency band without requiring the right of use, thus, the communication cost is low. Due to the small transmitting angle of infrared, the security of IrDA transmitting is strong. IrDA technology uses line of sight transmission to transmit data. Its communication can only be carried out between two devices, and the devices need to be aligned and barrier free. IrDA technology is widely used in PDA, mobile phones, portable computers, printers and other small mobile devices.

4.   ZigBee

ZigBee is a kind of short-range wireless communication technology similar to Bluetooth and based on LAN protocol with low power loss. IEEE 802.15.4 standard is its basic protocol. ZigBee works in a flexible frequency band. It cannot only use the same 2.4 GHz frequency band as Bluetooth, but also use 868 MHz (Europe) and 915 MHz (USA) frequency bands, all of which are unlicensed. The power loss of ZigBee is very low. The transmitting power is only 1mW. It can only rely on the battery to supply energy. Under the same power (two AA batteries), ZigBee can keep the device running for six months to two years, while Bluetooth can only keep for a few weeks, WiFi can only keep for a few hours. The cost of ZigBee devices is also lower than that of Bluetooth, mainly because ZigBee does not need to pay patent fees and the original cost of modules is lower.

ZigBee uses frequency hopping technology for data transmission. Its data transmission distance increases with the decrease of transmission rate. The data transmission rate is limited to 10–250 kb/s, and its basic rate is 250 kb/s. When the transmission rate drops to 28 kb/s, the data transmission distance is extended to 134 m, and the reliability of data transmission is improved. Therefore, ZigBee is suitable for low rate transmission. ZigBee can communicate with each other in two ways. It cannot only send control instructions to the device, but also receive feedback information from the execution status and related data of the device. ZigBee uses collision avoidance mechanism, and it sets aside a special time gap in advance for the communication services that need to keep the bandwidth unchanged, so as to avoid competition and conflict in data transmission.

ZigBee network structure is honeycomb type, and can be self-networked. Each node module can establish contact, and carry out data transmission. The network composition is flexible, at the same time, it can ensure the stability of the network. ZigBee can almost be considered no data pack drop. Compared with Bluetooth, ZigBee is more suitable for supporting entertainment, e-commerce, research devices and home automation applications. Each ZigBee network can support up to 255 devices, including 254 slave devices and one master device. There can be multiple ZigBee networks in the same monitoring range. ZigBee adopts AES-128 advanced encryption algorithm, supporting authentication authority and authentication authority. Its strictness is 12 times that of bank card encryption technology.

ZigBee is widely used in many fields, such as home and building network, industrial control, commerce, public places, agricultural control, medical treatment, etc., such as temperature control, automatic lighting control, curtain automatic control, gas metering control, remote control of household appliances, automatic control of various sensors and monitors, intelligent tags, environmental information collection, medical treatment, instrument sensor control, etc.

5.   NFC (Near Field Communication)

NFC technology is a kind of short-range wireless communication technology with high radio frequency. NFC integrates inductive card reader, inductive card and P2P functions in IC. NFC can achieve identification and information interaction with compatible devices in close range. NFC data transmission rate is divided into 106, 212 or 424 kb/s, and its working frequency is 13.56 MHz. There are two modes of NFC data reading: active and passive.

NFC uses the electromagnetic induction coupling method in the wireless frequency part of the spectrum to complete the data transmission, which is compatible with the existing contactless smart card. Different from other wireless communication technologies, NFC applies signal attenuation technology. NFC signal transmission distance is shorter than RFID, and its bandwidth is wider. NFC energy loss is smaller, can read data without power. Because NFC is a kind of short-range data transmission communication mode, its equipment must be very close, thus, its security is high. At the same time, with reliable identity authentication, it can provide safe and reliable interactive communication and data sharing between devices. NFC does not need manual setting, and its devices can be connected automatically.

NFC has three working modes: card mode, point-to-point mode and read/write mode, which can simplify the Bluetooth connection process. NFC can be used in access control system, real-time booking, e-commerce, mobile payment and other fields. NFC, IrDA, Bluetooth data transmission methods are non-contact. They have different technical characteristics, which can be applied in different communication occasions.

6.   UWB (Ultra Wide Band)

UWB technology is a carrier free communication technology which uses non sinusoidal narrow pulse with very small time interval to achieve information transmission. When wireless communication system transmits information, it needs to continuously transmit carrier signal, which will consume part of the energy. UWB transmits data by intermittent pulse instead of carrier, that is to say, the data is directly sent by 0/1. Only when it needs to transmit data, the pulse signal will be sent out. The duration of UWB pulse is very short, generally within 0.25 ns. The duty cycle is very low, the system transmit power is very small, thus, the system communication power consumption can be reduced to a very low level. In high-speed data transmission, the system power consumption is only a few hundred to tens of milliwatts.

Radio signal is usually with multipath propagation. UWB uses ultra wideband radio to complete the communication function. The radio wave is a single cycle pulse with very short duration and very low duty cycle. UWB has very strong multipath resolution and is easy to achieve accurate positioning.

The structure of UWB system is simple and easy to construct. After receiving the signal, the energy of UWB system will be restored. The spread spectrum gain will be generated in the process of signal de spread spectrum. The bandwidth of UWB is more than 1 GHz. When transmitting signals, a small radio pulse signal is diffused and distributed in a wide frequency band. The capacity of UWB system is very large. It can be paralleled with narrowband communication system without being affected by each other, thus, the anti-interference ability of UWB is very strong. UWB data transmission rate is large, which can reach tens of Mb/s to hundreds of Mb/s. Its transmission rate is limited by the transmission power.

UWB uses time hopping spread spectrum technology to disperse the signal in a very wide frequency band. It is not easy to detect the signal. The receiver can analyze the transmitted data only when it has known the spread spectrum code of the transmitter. For the general communication system, UWB signal can be considered as white noise signal. But in most cases, the power spectral density of UWB signal is lower than that of natural electronic noise. It is difficult to separate the pulse signal from the electronic noise. Especially after pseudo randomization of pulse parameters by coding, the pulse detection will be more difficult.

UWB technology can be fully realized by digitization in engineering. This process only needs a mathematical form to generate pulses, which can be debugged. UWB circuit can be integrated into a chip, thus reducing the cost of equipment.

There are many advantages of UWB: the structure of UWB system is relatively simple; the response to channel fading is slow; the security of data communication is strong; the transmitted signal has low power spectral density; and the positioning accuracy can reach centimeter level. These advantages enable UWB to be used both for military and civil. In the military aspect, it is mainly used in a variety of detection radars, military radio communication systems, UAV/UGV data links of radio stations and detection of buried military targets. In the civil aspect, UWB is suitable for short-range digital audio/video wireless links/access, various civil sensors, civil wireless data communication systems, etc.

7.   DECT (Digital Enhanced Cordless Technology)

DECT is a standard determined by ETSI. It is an open digital communication standard which is constantly improving. DECT can support voice and data communication services with high quality and low data transmission delay. In the process of signal transmission, the environmental change and other factors will cause signal switching. For example, the user movement may cause signal handoff, and the change of channel characteristics in a specific area will also cause corresponding carrier or time slot switching. DECT can ensure complete automatic switching without signal loss in these cases. DECT is a low-power system. Its channel allocation is dynamic, and there is no need for complex frequency planning. It can provide high traffic density and low cost. DECT has high security, with a relatively complete identity authentication mechanism. DECT has good information confidentiality, and is easy to install.

DECT is mainly used in home cordless telephone, commercial cordless communication system, wireless local loop, GSM/DECT integrated system, data service and multimedia service.

2.2.2 Communication Protocols of IOT

General networking communication protocol enables data sharing and interconnection between different devices, and integrates the system commonness of similar IOT equipment manufacturers. The protocol redefines the function communication protocol. It can share information between different systems, read/write data, and complete system control and other functions. Communication protocol is the rule and agreement that both sides of communication must follow to achieve information transmission or obtain corresponding services.

Communication protocol of IOT is divided into access protocol and communication protocol. Access protocol is the networking and communication protocol between subnet devices. Communication protocol is mainly the device communication protocol running on the traditional Internet TCP/IP protocol, which helps devices to exchange data and communicate through the Internet.

The communication network of IOT includes Ethernet, WiFi, 6LoWPAN (IPv6 low speed wireless version), ZigBee, Bluetooth, GSM, GPRS, 3G/4G/5G and other networks, and each communication application protocol has a certain scope of application. AMQP, JMS and REST/HTTP are all working protocols under Ethernet. CoAP is specially developed for resource limited devices (such as WiFi, Bluetooth, etc.). DDS and MQTT are highly compatible protocols.

The communication architecture of the IOT is established on the basis of the traditional Internet architecture. The protocol commonly used in the Internet is TCP/IP protocol. TCP/IP protocol is a set of protocols. HTTP protocol belongs to TCP/IP protocol. Because HTTP protocol has the advantages of low development cost and high degree of openness, it has a wide range of applications. Therefore, most of the manufacturers in the establishment of the IOT system is based on the HTTP protocol for development and research, including the physical web project led by Google, hoping to build the IOT protocol standard on the basis of traditional web technology.

HTTP protocol is a typical CS communication mode. The client initiates the connection request actively and requests XML or JSON data from the server. The original goal of HTTP protocol is to be able to use web browser to browse network information. At present, HTTP protocol is widely used in PC, mobile phone, Pad and other terminals, but HTTP protocol is not suitable for the IOT.

HTTP protocol requires the device to actively send data to the server, but it is difficult for the server to actively push data to the device. Therefore, HTTP protocol is only suitable for one-way data collection. For frequent operations, it needs to be completed through the device periodically and actively pulling data, which leads to the weakening of its implementation cost and real-time performance. HTTP is a clear text protocol. It requires high security of IOT. There are many kinds of devices in the IOT. It is difficult to implement HTTP protocol for devices with limited computing power and storage resources.

1. REST (Representational State Transfer)/HTTP

Rest is a kind of software architecture developed on the basis of HTTP protocol for network applications, rather than a standard. REST can reduce the complexity of R&D and improve the scalability of the system. REST architecture contains a set of constraints, that is, the system model is in the form of client/server. The component is unknown to the component outside the middle layer. The client and server are stateless before interaction. The client improves the data transmission performance by caching data. There is a unified interface to transmit state information between the client and the server. An application or design that meets all constraints is RESTful.

In the REST system, the server does not save any information about the user, thus, the user needs to provide enough information when sending a request every time. REST system needs appropriate cache to achieve loose coupling between client and server, reduce data transmission delay and improve communication performance. On the server side, the state and function of the application program can be divided into various resources to be exposed to the client. Each resource gets its address through the uniform resource identifier, and the address is unique. All resources share a unified interface to transmit information between the client and the server.

2. CoAP (Constrained Application Protocol)

Due to the limited computing power and storage space of many devices in the IOT, it is not appropriate to apply the traditional HTTP protocol in the IOT. CoAP is an application layer protocol based on REST architecture proposed by IETF. It is suitable for IP network communication with limited resources.

CoAP protocol defines four types of messages, which are to confirm (CON), not to confirm (NON), acknowledge (ACK) and reset (RST). This protocol takes message as the carrier of data communication. The data communication between devices is completed by exchanging network messages. In order to obtain the information resources of the server from the client, CoAP supports four message request methods: GET, PUT, POST and DELETE. The related operations of CoAP server cloud device resources are completed through the request and response mechanism. The CoAP protocol is based on two-way communication technology, which can achieve asynchronous communication. CoAP client and CoAP server can send message requests to each other independently. CoAP protocol is based on user datagram UDP protocol, rather than traditional TCP protocol, which can achieve multicast function with low cost. CoAP protocol has high reliability of data transmission. It can send requests to multiple devices at the same time. Its protocol packet is very small. The minimum length is only 4B, and the typical request header is 10–20B. In CoAP protocol, there is a resource discovery format of discovering device message list or device broadcasting its own message to service directory. In CoRE, the message path is described in the format of "/.well-known/core". CoAP can cache the description of resources and optimize its data processing performance. CoAP protocol is suitable for low-power IOT.

3.   MQTT (Message Queuing Telemetry Transport)

MQTT (low bandwidth) is a more suitable instant messaging protocol for the IOT. MQTT protocol uses publish/subscribe mode and supports all platforms. It can connect almost all IOT terminals and physical devices, and can be used as the communication protocol of sensors and brakes.

Considering the difference of computing performance of different devices, MQTT protocol adopts binary encoding/decoding format, and the encoding/decoding format is easy to develop and implement. The minimum packet contains only 2 bytes, which is suitable for low power consumption and low rate network. MQTT protocol runs on TCP protocol and supports TLS (TCP + SSL) protocol. Because all data communication passes through the cloud, it improves the network security.

MQTT protocol is a communication protocol based on cloud platform for data transmission and monitoring of remote devices. It mainly provides services for remote sensors and control devices with limited computing power and working in low bandwidth and unreliable network. It mainly has the following characteristics.

(1)   Using publish/subscribe message mode. One to many message publishing mode can be realized to decouple the application.
(2)   Using TCP/IP protocol to provide network connection.
(3)   When transmitting information data, the load content is shielded.
(4)   It has a relatively perfect QoS mechanism, and selects the message delivery mode according to the actual application requirements.

At most once: this mode completely relies on TCP/IP network to complete the release of messages. This way of message delivery will lead to the loss or repetition of messages.

At least once: this mode ensures messages arrival, but message duplication may occur.

Only once: this mode ensures that the message arrives once.

(5) The device sending the "last words" is interrupted abnormally, and the "last words mechanism" and "will mechanism" are used to notify other devices.

(6) Small transmission, protocol switching minimization, reduce network traffic.

MQTT protocol is a lightweight open communication protocol, which is easy to implement and widely used. MQTT protocol has been widely used in sensors using satellite link communication, smart home and some small devices.

4. DDS (Data Distribution Service for Real-Time Systems)

DDS is a middleware specification for distributed real-time communication, which uses publish/subscribe architecture. With data as the core, DDS provides a wide range of QoS guarantee approaches, with as many as 21 kinds of QoS strategies. DDS has strong real-time performance, can well support data distribution and device control, and has high data distribution efficiency. It can simultaneously distribute millions of messages to many devices within seconds. DDS ensures the real-time, high efficiency and flexibility of data distribution, and can meet the application requirements of various distributed real-time communication. DDS can be interoperable, that is, the application systems developed by different manufacturers and platforms can be directly interconnected. At the same time, DDS can be controlled across platforms, allowing the use of a variety of operating systems and hardware platforms, as well as a variety of underlying physical communication protocols. It has a full life cycle of "simulation → test → real installation".

DDS is suitable for distributed, high reliability, real-time transmission occasions. DDS is widely used in national defense, civil aviation, industrial control and other fields. It is a standard solution to publish and subscribe data in distributed real-time system.

5. AMQP (Advanced Message Queuing Protocol)

AMQP is the application layer standard protocol of unified messaging service and can support multiple message interaction architectures. No matter whether the client, middleware product or development language are the same or not, the client and message middleware based on this protocol can deliver messages to each other. AMQP is a binary protocol, which has the characteristics of multi-channel, negotiation, asynchronous, secure, cross platform, neutral, efficient and so on.

AMQP can be divided into three layers: model layer, session layer and transport layer. The model layer defines a set of instructions according to its functions, through which various applications can complete their tasks. The instruction transmits information from the device to the server through the session layer, and then feeds back the processing result of the server to the device. The transport layer can achieve frame processing, channel reuse, error detection and data description.

In the IOT, AMQP protocol is mainly applied to the communication and analysis between mobile handheld devices and background data center.

6.   XMPP (Extensible Messaging and Presence Protocol)

XMPP (instant messaging) is an extensible communication and presentation protocol, which is based on Jabber open protocol and is a subset protocol based on Standard General Markup Language XML. XMPP follows its flexibility and development in XML environment, and is suitable for real-time communication and representation of service class and streaming transmission of XML data element in demand response service. XMPP has a strong application scalability. After expansion, XMPP can meet the requirements of users by transmitting relevant extended information, and can develop related applications at the top of XMPP. XMPP includes the software protocol applied to the server, which can lay the foundation for the mutual communication between servers.

XMPP includes client, server and gateway. Any two can communicate with each other. The server records the data information of the client, controls the connection, and exchanges the information. The gateway can connect and interact with SMS, MSN, ICQ and other heterogeneous instant messaging systems. A client is associated with a server through TCP/IP protocol, and then transmits XML on the server to form a basic network form.

XMPP is used for instant messaging (IM) and online on-site monitoring. Even if the user's operation system and browser are different, XMPP protocol can also allow network users to send real-time messages to anyone else in the network. XMPP protocol is suitable for applications that need instant communication, as well as network management, collaborative work, file sharing, games, remote system monitoring and other fields.

XMPP is a distributed network structure based on client/server mode. XMPP client structure is simple, most of the work is running on the server. XMPP protocol is free, open and easy to understand. It has been implemented in many aspects, such as client, server, component and source code library. XMPP protocol is a long XML file with single encoding, in which the binary data cannot be modified. It needs to be applied in combination with external HTTP protocol. XMPP protocol also provides a Base64 encoding mode that can transmit long identification information in HTTP environment. XMPP has been widely used in the application of Internet timely communication. Compared with HTTP, XMPP is more suitable for the IOT system in terms of communication process. Developers do not need to know the communication process of a device in detail. Therefore, the development cost is relatively low. However, the security of HTTP protocol and the consumption of computing resources have not been fundamentally solved.

7.   JMS (Java Message Service)

JMS is the Java message service application program interface, which is located on the Java platform. JMS is a message oriented middleware (MOM) API. It has nothing to do with the specific application platform. JMS can transfer messages between two applications or in a distributed system. It supports synchronous and asynchronous message processing.

JMS includes two message modes: peer-to-peer and publisher/subscriber.

In the peer-to-peer model, a sender transmits information to a designated queue, and a receiver receives information from the sender from the designated queue. During the receiving end processing the information, the sending end does not need to be in the running state. The receiving end does not need to be in the running state during the information transmission.

The publisher/subscriber mode supports publishing messages to a specified message body. Some subscribers may be interested in this message, and the sender and subscriber do not know each other, but the publisher and subscriber are time-dependent. Publishers need to create a subscription that can be purchased by users. The subscriber needs to maintain its own activity to receive messages. If the subscriber decides to subscribe for a long time, the messages transmitted when the subscriber is not connected will be sent back when the subscriber establishes a connection.

JMS uses Java language to separate application layer and data transmission layer. According to the provider information contained in JNDI, the same set of Java classes can be connected to different JMS providers. The message transmission between JMS clients is achieved by message mediator or router. The message is composed of two parts: the header containing interactive information and related message metadata and the message body containing application data and payload. According to the different load types carried by the transmission message, the message can be divided into simple text, attribute set, byte stream, serializable object, original value stream and no payload message.

In summary, the results of the comparison of the IOT protocols are shown in Table 2.1.

DDS, MQTT, AMQP, XMPP, JMS, REST, CoAP protocols are widely used in the Internet, and each protocol can be implemented by a variety of codes. Theoretically, they are IOT protocols that support real-time publish/subscribe. However, when building a specific IOT system, it is necessary to select the appropriate protocol according to the actual communication requirements.

DDS, MQTT, AMQP, and JMS all adopt publish/subscribe mode. Publish/subscribe framework is more suitable for communication in the IOT environment. Because this framework has the characteristics of service self discovery, dynamic expansion and event filtering. It solves the problems of rapid access to data sources, things joining and exiting, message subscription and so on in the application layer of the IOT system. This framework achieves loose coupling connection and synchronization of things in temporal and spatial. Quality of service (QoS) is very important in the communication of IOT. By using mobile service strategy, DDS protocol can effectively control the use of network bandwidth, memory space and other resources. At the same time, it can also improve the real-time performance, reliability and prolong the data lifetime.

Take the communication protocol used in the construction of smart home IOT as an example: XMPP protocol can be used to control the smart light switch in smart home. DDS protocol can be used for the power supply of smart home and the engine monitoring of power plant. MQTT protocol can be used to inspect and maintain the lines in the process of power transmission. AMQP protocol can be used to obtain

Table 2.1 Comparison of IOT protocols

Characteristic	REST/HTTP	CoAP	MQTT	DDS	AMQP	XMPP	JMS
Abstract	Request/Reply	Request/Reply	Pub/Sub	Pub/Sub	Pub/Sub	NA	Pub/Sub
Architecture style	P2P	P2P	Agent	Global data space	P2P or proxy	NA	Agent
QoS	Guaranteed by TCP	Confirmation or non confirmation message	3 types	22 types	3 types	NA	3 types
Interoperability	Yes	Yes	Part	Yes	Yes	NA	No
Performance	100 req/s	100 req/s	1000 msg/s/sub	100,000 msg/s/sub	100 msg/s/sub	NA	1000 msg/s/sub
Hard real time	No	No	No	Yes	No	No	No
Transport layer	TCP	UDP	TCP	The default is UDP, which is also supported by TCP	TCP	TCP	Not specified, usually TCP
Subscription control	NA	Support multicast address	Topic subscription with hierarchical matching	Topic subscription for message filtering	Queue and information filtering	NA	Topic subscription for message filtering
Code	Plain text	Binary	Binary	Binary	Binary	XML text	Binary
Dynamic discovery	No	Yes	No	Yes	No	NA	No
Security	Generally based on SSL and TLS		Simple user name/password authentication, SSL data encryption	Provider support, generally based on SSL and TLS	SASL authentication, TLS data encryption	TLS data encryption	Provider support, generally based on SSL and TLS, JAAS AF support

household appliances energy consumption information, which can be transmitted to the cloud or the home gateway for data analysis. Finally, if users want to upload the energy consumption query results to the Internet, they can use REST/HTTP to open the API service.

2.3   Edge Networking

2.3.1   Edge Computing

The basic function assumption of the "Cloud-Network-End" architecture of the IOT is that the cloud is responsible for data processing. Therefore, the cloud needs to be equipped with a powerful data center. The network is responsible for data transmission. The data collected from each node of the IOT is transmitted to the cloud. The cloud completes the analysis and processing of the data, and transmits the results to the terminal after decision-making. In this hypothetical architecture, the cloud carries out intelligent analysis and calculation of data, and the terminal is responsible for data collection and implementation of decision results.

We assume that the difficulty of the model comes from the data on the one hand and the network delay on the other. There are many data in the IOT. The IOT transmits data to the cloud through the wireless network. If the nodes of the IOT upload all the original data without processing, the network cannot bear the bandwidth demand, which will lead to the explosion of bandwidth demand. At the same time, if the unprocessed data information is completely uploaded to the cloud, the wireless transmission module of the terminal node of the model must be able to meet the needs of high-speed transmission. Therefore, the wireless transmission module will produce a large power loss. For the hypothetical model, the power loss requirement of the model is low. Network delay has a great impact on some applications. In order to solve the data processing problem and network delay problem of hypothesis model, scholars propose edge computing technology.

Edge computing is an open platform that combines networking, computing, storage and application capabilities at the edge side of the network near the object or data starting end. It implements edge intelligent services nearby and realizes many functions, such as fast and convenient connection of applications, real-time task processing, optimization of data processing, protection of security and privacy. Edge computing is a mesh network that can process and store key data locally, and transmit the received data to the central data center or cloud repository. It can process and analyze the data close to the data source. The intelligent device itself is a data center, and it does not need to transmit data to the cloud for processing through the network. Its basic analysis and processing are carried out on the device, thus, the network delay is reduced and the real-time response can be carried out. At the same time, due to the dispersion of devices, data processing on different devices can reduce network traffic, bandwidth requirements and network power consumption. The cloud can do further evaluation and analysis for the processed data.

Edge computing adopts the modern communication network, with cloud computing as the core, perceives through the huge terminal, optimizes the resource allocation, and makes computing, storage, transmission, application and other services more intelligent. Edge computing has the ability of resource scheduling with complementary advantages and deep collaboration. It is a new computing model integrating "cloud, network, end and intelligence".

With the development of IOT and information technology, intelligent edge computing has emerged. Intelligent edge computing enables edge devices of IOT to collect, process, communicate and analyze information. Therefore, the edge sensor can judge the sensing data by itself, instead of continuously transferring all the sensing data to the data center for processing. Only when the sensor collects abnormal data, it will be uploaded to the data center for decision-making.

Intelligent edge computing can also conduct large-scale security settings, arrangements and management of edge devices through the cloud, and intelligently allocate the cloud and edge devices according to the device type and application environment. Different from the application of cloud computing in non real-time, long period data and business decision-making, edge computing is suitable for real-time, short period data and local decision-making. The number of IOT nodes is large, the data transmission delay between them is also large. The need for efficient equipment management, the higher security level requirements, the demand for edge computing is more urgent.

For the IOT with more network nodes, a large number of key data will be generated in its network nodes. The IOT needs to quickly analyze and apply the data, which requires the method of edge computing. Due to limited conditions, the cost of data transmission to the data center or cloud platform is relatively high, and the delay time is also relatively long. Edge computing is carried out at the edge of the network around the object or at the starting end of the data. When transmitting data to the cloud, some data are preliminarily processed at the edge node, which shortens the response time of the device and the amount of data transmitted from the device to the cloud. At the same time, the filtering and analysis of data and information are realized at the edge node, which can quickly extract specific data and send it to the cloud platform. In order to maintain the efficiency of device management and process information more efficiently. When the IOT needs a higher level of security, edge computing is used to encrypt, authenticate and protect data at the edge, and the security mechanism is embedded in the edge in a distributed way. Various resources such as network, computing and storage are closer to the end user. Therefore, users can use the edge computing to encrypt, authenticate and protect data. Edge computing can obtain the nearest edge intelligent services to meet the needs of fast connection, real-time update, data optimization, intelligent applications and so on. Because edge computing is suitable for real-time and short period data analysis, it can well support the real-time intelligent processing and execution of local business. Compared with real-time performance, storage and bandwidth cost, edge computing has more advantages than cloud computing.

By placing storage devices and computing resources near the data source, the delay can be greatly reduced and the cloud bandwidth required for data transmission can be reduced. Because edge computing can reduce the amount of data transmission on the network, it can improve the network security. However, with the growth of the number of IOT devices and infrastructure levels (including edge servers), there are more attachable points. Every new node increases the security threat of the cloud, and also increases the way to penetrate the core of the network.

The application advantages of IOT edge computing include almost zero delay, less network load, greater flexibility, higher security and reliability, and lower data management cost.

1. Almost zero delay

In the traditional mobile cloud computing technology, the client access to the cloud through the wide area network (WAN). The communication delay of WAN is relatively long and has a strong uncontrollability. Excessive delay and jitter seriously affect the control effect and even cannot complete the task. The time interval between data acquisition, processing and response of edge computing is almost zero, which is very important for IOT devices to complete delay sensitive tasks.

Google estimates that its autopilot car generates about 1 GB data per second. It requires the network to process large amounts of data quickly so that the car can maintain the correct route and avoid collision. If these data are collected and transmitted to the cloud, processed in the cloud, the results will be fed back to the vehicle from the cloud. Although the whole process can be completed in a few seconds, in fact, the car may have collided. Therefore, the best solution is to use edge computing to analyze the data collected by the sensor itself, and then send the processed data to the cloud for subsequent analysis.

Edge computing is also critical in the medical industry. For example, if medical devices are connected with heart rate monitors or pacemakers, slight delay may affect the lives of patients. In the industrial field, the industrial system with a large number of sensors uses edge computing to reduce the data delay, which promotes the development of industrial automation. At present, the robot/UAV has been widely used in military operations, disaster rescue, social services and other aspects. To ensure that the robot/UAV can complete the control requirements in time, it needs to process the data quickly and feedback the results, which can be completed by using edge computing.

2. Less network load

Cisco estimates that by 2020, the amount of data processed by IOT devices will reach 7.5 ZB (1 ZB = 1000 GB). The large amount of data on the Internet will increase the possibility of network congestion, especially in areas with weak networks. Using the edge computing method, most of the traffic load will be processed on the edge devices, instead of sending all the data to the cloud through the network. Therefore, the network congestion can be effectively improved.

3. Greater flexibility

Connecting devices in the IOT can enhance their connection flexibility through the distributed architecture provided by edge computing. A single virtual machine failure in the cloud will affect all IOT devices connected to the network. Through the distributed architecture of edge computing, even if one of the devices fails, it will not affect other devices, and the rest of the devices will still maintain normal operation.

4.   Higher security and reliability

The security and reliability of data transmission is very important in mobile cloud computing. In the IOT, computers, mobile phones and other application devices are connected though wireless methods, and then transmit data to each other through the network. Data transmission is multi hop between nodes until it reaches the destination, that is, it needs to connect to cloud services through ad hoc network. In the process of data transmission, information is easy to leak or be maliciously modified. The more complex the network and the more hops the data passes through, the less secure and reliable it is. In many cases, users will automatically request access to cloud services. Remote access may have serious security risks. Edge computing processes data at the edge of the network, which reduces the amount of data sent through the network and helps to reduce the possibility of information leakage in transmission.

5.   Lower data management cost

The use of edge computing only stores the processed data in the cloud, thus, it can reduce the cost of cloud storage. Because the amount of data is relatively small, it is conducive to the efficient implementation of management. Only when more in-depth analysis is needed, the summary data will be sent to the cloud for analysis and decision-making.

2.3.2 Networking of IOT

The current challenges of the IOT can be summarized as follows: the rapid growth of wide area distributed mobile edge data leads to the increase of network equipment communication load of data collection, processing and service. Most of the edge nodes are uncertain and dynamic, which contradicts with the characteristics of timeliness and regionality of edge data. It is difficult to manage, process and character extraction of the data.

In view of the above problems, we can study how to sample and process the massive data of the wide area distributed network, and apply it to the mobile edge node networking, so that the network can make full use of a large number of idle resources of the mobile edge, and assist the data center to improve the information processing ability.

Starting from Internet architecture and wireless networking, researchers study networking technology, including content delivery network (CDN), software defined network (SDN), network functions virtualization (NVF), delay tolerant network (DTN), wireless mesh network (WMN), mobile edge computing (MEC), etc.

1. CDN

CDN is a kind of network content service system, which is built on the basis of IP network. The system uses intelligent virtual network, according to its network traffic, the association status of each node, load status, the distance to the user end and response time and other related information resources, timely sends the user's instructions to the nearest service node. The user can get the relevant demand information quickly in the shortest time. Using CDN can effectively improve the information congestion of the Internet, and improve the response speed of users visiting the website.

The content distribution and service functions of CDN meet the requirements of efficiency, quality and content order of content access and application. They have the characteristics of high network service quality, high efficiency and distinct network order.

CDN achieves multi-point redundancy in architecture. Even if there is a node that fails due to a special event, it can be automatically transmitted to the other healthy nodes to respond to the website visit. CDN can cover most of the lines in China, without considering how many servers need to be set in the network and the hosting of servers, the cost of new bandwidth, the synchronization of multiple server images, and more management and maintenance technicians.

Using CDN, no matter in any time and place, through any network operator, users can quickly access the website. Using CDN can reduce the initial construction and operation costs of various server virtual host bandwidth, and the website traffic, consulting volume, customer volume and order volume will be improved through CDN.

2. SDN

SDN is a kind of network innovation architecture, which takes OpenFlow technology as the core, separates the control plane and data plane of network device. SDN achieves the purpose of mobile control of network traffic, and provides a good platform for core network and its application innovation.

Traditional network devices such as switches and routers are set and managed by device manufacturers. SDN can separate network control from physical network topology to solve the problem that network architecture is subject to hardware parameters, so that enterprises can modify, adjust, expand or upgrade the whole network structure to meet their needs. When modifying the network architecture, there is no need to replace the basic network device hardware, which can not only reduce the construction cost, but also shorten the network structure iteration cycle.

3. NFV

NFV applies virtualization technology to divide the functions of network nodes into different functional blocks, and operates them with software, which is a network architecture not limited by hardware architecture.

The core function of NFV is virtual network function, which should be combined with application, work flow, integrated and adjustable infrastructure and software. NFV is applied in routing, customer terminal, IMS, mobile core, security, policy and so on. NFV technology is not for the purpose of device customization, but hopes to achieve the relevant network services on the standard server. NFV integrates the network device types into standard server, switch and storage device, so as to simplify the open network elements.

The important decisions to deploy NFV include setting up cloud hosting mode, selecting appropriate platform to achieve network optimization, providing services and resources to promote operation integration, and building flexible and loosely coupled data and process architecture.

4. DTN

DTN is a special kind of network, in which it is difficult to establish an end-to-end path. The delay of message propagation in the network makes DTN unable to use the traditional TCP/IP protocol.

DTN network has some basic assumptions and characteristics different from traditional Internet protocol. DTN has a long network delay time, and the general information transmission time is more than minutes. The instability of network connection is high, and the connection between network nodes is easily affected by node movement, connection failure, disturbance and other factors, which is often intermittent connection. The transmission rate of DTN link is asymmetric, and its data transmission has a high bit error rate and error rate, and the loss of data packets probability is high.

DTN adds a bundle layer to the transport layer to form an overlay network architecture. DTN architecture can form an interconnected heterogeneous network. This network can allow long delay and connection interruption, which liberates the limitation of some assumptions of TCP/IP protocol.

5. WMN

Ad Hoc network is a mobile network without wired infrastructure support. WMN is a special form of this network. It combines the advantages of WLAN and Ad Hoc network, and can solve the bottleneck problem of transmission line from broadband access end to end user. The topology of WMN is grid format, and the network is composed of IAP/AP, WR and Client.

AP is the wireless access point. An AP can be effective within tens of meters or even hundreds of meters. It can connect multiple wireless routers within the effective range. AP is equivalent to the switch in the wireless network. Its main function is to connect the wireless network to the core network, and connect the wireless clients interconnected with the wireless router into one, so that the terminal device can share the information resources of the core network through AP after installing the wireless network card. On the basis of AP, the intelligent access point IAP is constructed by integrating the routing function of Ad Hoc. AP/IAP can realize the management and control of wireless access network. By adding the intelligent performance of traditional switches into AP/IAP, the construction cost of core network can be reduced and the network scalability can be enhanced.

The lower layer of IAP is equipped with wireless router (WR), which provides packet routing function and data forwarding function for mobile terminal devices. Users can use IAP to download resources and update wireless broadcast software. According to the available nodes at this time, which router can forward packets is decided in real time, that is, dynamic routing. In the network architecture of WMN, the main function of wireless router is to flexibly expand the range of information transmission between mobile terminal device and access point.

The client can be used as both host and router. This is because the node can not only be used as an application related to host work, but also as a routing protocol related to router work. It plays an important role in routing search, routing maintenance and other routing work.

WMN has two types of implementation modes, namely infrastructure grid mode and end user grid mode.

In the infrastructure grid mode, access points are connected with terminal users. The data transmission and relay functions of wireless routers are mainly used to realize the connection between mobile terminals and IAP. At the same time, the terminal user optimizes the communication path selection of the destination node by using routing and control functions. Mobile terminals can connect with other networks through IAP and access wireless broadband. This mode saves the system cost, and improves the network coverage and network reliability.

In the end user grid mode, the user carries the wireless transceiver and uses the wireless channel connection to construct a point-to-point arbitrary grid topology network. Because the nodes can change their positions at will, the network topology changes accordingly. The end-user grid model is essentially an Ad Hoc network that maintains the data interaction environment without network infrastructure or using the existing network infrastructure. In the limited range of wireless terminal communication, the user terminal which cannot directly interact with each other can use the packet forwarding function of other terminals to complete its information interaction. The terminal device can complete the work independently, and the mobile terminal can move at a high speed and quickly build a broadband network.

The advantages of infrastructure network and end user network are complementary. Therefore, WMN which supports the two modes at the same time has a wide range of wireless communication. The end user cannot only complete the wireless broadband access, connect with other networks, directly transmit data with other users, but also act as an intermediate router to forward the data of other nodes.

Compared with the traditional wireless network, WMN has stronger reliability, and can avoid conflict. It has simple link, large network coverage, flexible networking and low cost. WMN has conflict protection mechanism. In order to reduce the interference between links, the collision link can be marked and the angle between links is obtuse. The wireless link length of WMN is generally short, which can not only save the cost, but also reduce the data transmission distance and network performance. The interaction between RF signals of different systems and the disturbance of the system itself decrease with the decrease of transmit power. WMN reduces the transmission power, which can not only reduce the system disturbance, but also contribute to simplify the link. WMN introduces WR and IAP, which enables end users to access the network anytime and anywhere or connect with other nodes, expands the range of access points, improves the utilization of spectrum and expands the system capacity. In view of the characteristics of WMN network with Ad Hoc network, we only need to add some WR and other wireless devices in a specific area to form a wireless broadband access network with existing facilities. WMN network routing path selection function makes the link interruption or local expansion, upgrade will not have a serious impact on the whole network work, thus, improving the feasibility and flexibility of the network. Its function is more powerful and perfect compared with the traditional network. The location of AP and WR in WMN network is basically fixed, which reduces the network resource overhead and reduces the initial cost of WMN network construction.

6. MEC

MEC is a system combining software and hardware. It constructs a carrier level network service environment with high performance, low data transmission delay and high bandwidth. By using mobile edge computing, it can quickly download various resources, services and applications of network terminals, so that consumers can have a continuous, efficient and fast network experience.

MEC integrates wireless network technology and Internet technology, and adds data calculation, storage, processing and other functions in its wireless network end. It introduces related applications through open platform, uses wireless API interface to connect open wireless network and business application server, and transmits information, and combines wireless network with work to build an intelligent wireless base station. MEC provides special services for different industries and improves the network efficiency. At the same time, the planning of mobile edge computing, especially the planning of geographical location, can shorten the delay and improve the bandwidth. MEC can provide more accurate services by collecting wireless network information and location information in real time.

According to the development of mobile edge networking, it can be divided into intelligent networking, semi intelligent networking and non intelligent networking. Intelligent networking firstly uses machine learning method and then according to the data features extracted independently to network. Semi intelligent networking manually extract features to network. Non intelligent networking is to network according to the fixed rules specified by human.

2.3.3 MEC Networking Scheme

Figure 2.14 shows the MEC server platform architecture proposed by ETSI (European Telecommunications Standards Institute). In order to achieve the virtualization

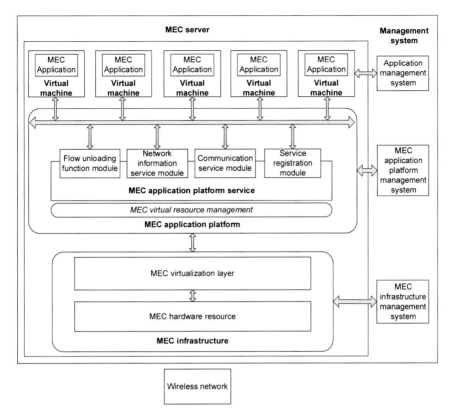

Fig. 2.14 MEC server platform architecture proposed by ETSI

of physical resources in the real world and implement the unified management of virtualized resources, MEC platform uses a hierarchical structure. Operators can control the application platform, application life cycle and other related functions through the management interface of the whole platform provided by MEC platform. According to the requirements of MEC, edge cloud platform provides infrastructure services, wireless network information services and traffic unloading services. Infrastructure services include communication services and service registration.

Using the communication service function of infrastructure services, information can be transmitted between applications running on the cloud platform or between applications and the cloud platform through a specific system interface. Communication service function can achieve decoupling operation of application, complete one to many information broadcast, and also can use effective protection mechanism to resist malicious applications.

Service registration refers to the list of service types, related interfaces and versions supported by the edge server, so that applications can find and locate the required service locations, make flexible arrangements for applications, and provide location sharing services for other applications.

MEC allows authenticated applications in the edge server to obtain the real-time information provided by the wireless network information service module. The real-time information contained in the wireless network information service module includes: data resources related to user device access, parameters related to user, statistical information, etc. Third party applications can process high-level information according to the network information.

The flow unloading module can control the flow of the wireless access network on the packet layer. The control order can be specified as that the application that passes the authentication will be controlled first. Flow unloading module plays an important role in balancing system flow and QoS.

1.   MEC networking hierarchy model

MEC adopts a layered network architecture to meet the needs of data calculation, storage and resource allocation in MEC network, which makes the network flexible and scalable. The layered structure mainly includes physical infrastructure layer, virtual resource layer, control layout layer and application layer.

1)   Physical infrastructure layer

The physical infrastructure layer is composed of network devices and IT devices. IT devices are mainly industrial class servers that can calculate and store resources. WiFi, which is composed of optical fiber subnet and wireless subnet, is usually combined with MEC network resources in access network. The physical infrastructure of optical subnet includes switch, gateway and optical link. The physical infrastructure of wireless subnet includes 3G/4G/5G base station and wireless access node.

2)   Virtual resource layer

Using virtualization technology to abstract the computing resources, storage resources and network resources of physical infrastructure layer, the virtual computing resources, virtual storage resources and virtual network resources form a virtual resource layer, which can flexibly manage the computing, storage and network resources. Figure 2.15 shows the MEC networking model.

3)   Control layout layer

The control layout layer consists of SDN controller and NFV controller, which can calculate and store the virtualization data of virtual resource layer. SDN controller and NFV controller provide a complete resource integration and management platform for the specific application work of the application layer, so that the virtual machine in the application layer can work independently. It does not rely on the information resources of the physical infrastructure layer, and extract the corresponding virtual resources according to the application requirements.

Due to the heterogeneous characteristics of optical fiber network and wireless network, traditional communication methods are inflexible, slow response and unable to carry out network virtualization. Therefore, traditional methods have limitations in

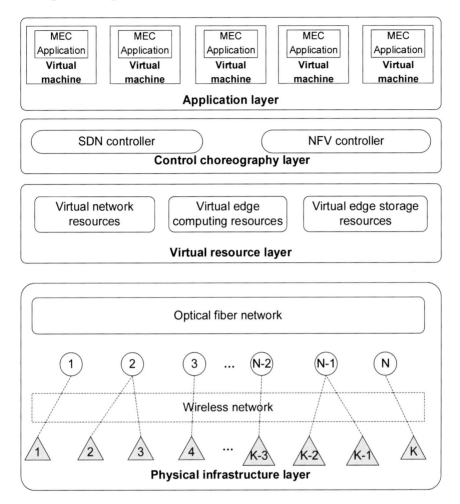

Fig. 2.15 MEC networking model

the management of optical fiber and wireless network. In order to solve the limitation of traditional methods to network management, SDN controller is used to build the network model to achieve the programmability of edge network and enhance the network flexibility brought by edge computing. At the same time, the introduction of NFV scheduling mechanism, the usage of general server instead of dedicated server equipment, reduce the cost of network construction. NFV controller can connect routers, firewalls, gateways, WAN accelerators and other physical devices to the data center according to the standard IT virtualization technology, making the devices also programmable. The network model constructed by SDN and NFV controller can integrate and process network resources, strengthen the management and control of the whole network. It carries out automatic network distribution, fault diagnosis and network maintenance, reduces network operation cost and improves network reliability.

4)   Application layer

Application layer combines computing, storage and network resources in the form of virtual machine. Third party applications and users can use application layer information. There are various services running in virtual machine on MEC platform. Different virtual machines are independent of each other. Information interaction and communication are completed through MEC platform.

At present, the network delay is long and the reliability is not strong. The MEC network hierarchy model introduces NFV controller and SDN controller through a special networking mode. It applies NFV scheduling mechanism to achieve effective virtual computing, storage and network resource management. The rapid arrangement and mechanical expansion of edge cloud is achieved. An independent virtual running environment for various applications working at the network edge is also provided. SDN controller can build edge network and process data directly according to specific rules at the edge. MEC networking mode can obtain the global view and complete the control and forwarding operations separately. Its routing and data forwarding mode does not take the node as the core, which can reduce the network delay and improve the network reliability.

2.   Key technologies of MEC networking model

The key technologies of MEC networking model include network function virtualization technology, software defined network technology and edge operating system. The work of data calculation and storage resource sorting management in MEC is achieved by network function virtualization technology. The software defined network technology can guarantee the programmability of edge network resources. The edge operating system makes the whole MEC platform scalable and can speed up the edge cloud distribution processing speed.

1)   Network function virtualization technology

The arrangement of mobile edge networks and devices needs to be flexible and scalable to meet the needs of users to the greatest extent. Different software and hardware from different equipment suppliers need to be integrated before application. In order to improve the flexibility and scalability of the platform, cloud technology and virtualization technology can be used. Therefore, in MEC, the key technologies of cloud computing capability and IT service environment also include cloud computing technology and virtualization technology.

Platform virtualization technology will extract independent logical structure in its hardware layer, so that resources can be flexibly applied. Virtualization technology has good automation performance and high flexibility in the field of network. Virtualization technology enables standardized procedure and products to work on a common server architecture. For MEC, standard service products work on the general server architecture at the edge of mobile network, and MEC needs to have cloud computing function to serve the third party. When virtualization technology is applied in MEC, multiple virtual machines can be placed on the same MEC platform. These virtual machines share hardware resources by effective control strategies. At the same time, network function virtualization technology can remove the coupling problem between the third-party application and software environment in virtual machine and the basic hardware resources. The universality and scalability of MEC server will be enhanced.

2) Software defined network technology

Software defined network (SDN) transmits network switches as packets controlled by the center, separating the control layer and forwarding layer of the network. SDN enables network devices to obtain a global view. SDN switches can be added to MEC servers to enable them to control the overall access network.

The network resources of different access networks need to be managed by different SDN networking schemes. Using the method of software defined central control, a two-tier wireless network can be constructed. The access node and the central controller jointly control the network. This structure can effectively balance the load, manage the interface, maximize the throughput and increase the utilization of wireless access network. SDN can extract the basic physical resources well, and ignore the information of different structures and characteristics between different networks. The application of SDN technology in the wireless access network can flexibly allocate and process the virtual network information according to the actual conditions. It achieves the efficient allocation of physical layer resources, realize the dynamic backhaul configuration and control the connections in the wireless network.

3) Edge operating system

Edge operating system is a solution to sort and control edge cluster. Manzalini et al. proposed an edge operating system software architecture that can work on a variety of operating systems. This architecture can upgrade the existing telecommunications infrastructure to a resource rich network service platform.

There are two types of nodes in the edge operating system: master node and ordinary node. The function of the master node is to provide the domain name server and each node with the required service registration information. Ordinary nodes obtain the data information of other registered nodes by establishing a connection with the master node and transmitting information, and communicate with this node. The data level interconnection between nodes is achieved through wired or virtual wireless links. Each node needs to interact with the registered master node, and update the running status of the associated physical hardware in time. The database of the master node can store a large amount of real-time data. According to the actual needs, corresponding functions are generated (or deleted) in the control edge area of the master node to respond. The master node should not only be connected with ordinary nodes, but also interact with high-level choreography layer and other regional master nodes to transfer information, so as to ensure that the service chain can be set across regions. The system database provides the relevant resources of the area where the infrastructure belongs.

In order to improve the virtual resource allocation rate, a common virtual resource pool is set between the master node and the ordinary node. After the master node issues the task command, if the virtual resources stored in the common node physical facilities to complete the task are not enough, the resources in the common virtual resource library can be used. Nodes adopt hierarchical recursion method to improve the scalability and scalability of the system. After virtualization, resources can be built into larger virtual resource entities to complete more complex computing tasks.

The edge operating system has an independent choreography layer. Service requests that need cross regional cooperation to complete tasks can be received and processed through the choreography layer. The database of the choreography layer will save and update the registration information of different master nodes in real time, split the requests in the service chain, and select the appropriate master node in the database to achieve cross regional end-to-end task assignment.

The hierarchical structure of edge operating system realizes the feature extraction and flexible configuration of physical resources such as computing, storage, network and so on. The node recursive structure of edge operating system ensures its scalability and scalability.

2.4 Sensors in IOT

The IOT can perceive object information through the perception layer. Sensors installed on various real objects such as power grids, water supply systems, railways, bridges and household appliances are connected through the network. According to the specified procedures and protocols, the communication between things is achieved. The IOT uses sensors, radio frequency identification, QR code and other devices or technologies to connect with the wireless network through the interface to achieve the intelligence of objects. Therefore, the IOT is based on sensors to operate things.

The information perception system composed of data acquisition layer and network layer is the key research object of IOT, and the wireless sensor network (WSN) plays a key role.

2.4.1 Sensor

1. Sensor concept

According to GB/T 7665-2005 National Standard for General Terms of Sensors, a sensor is defined as "a device that can sense a specified measurement and convert it into a usable signal according to a certain rule. It is usually composed of sensitive elements and conversion elements."

The function of the sensitive component in the sensor is to sense the measured signal and output the physical quantity signal which has a clear relationship with the measured signal. The physical quantity signal output by the sensitive component is used as the input signal of the conversion component and the electrical signal is output through conversion. The electrical signal output by the conversion component is amplified and modulated through the conversion circuit. The sensor supplies power to the conversion component and the conversion circuit through the auxiliary power supply.

2.   Sensors classification

According to the function of sensors, they can be divided into force sensor, liquid level sensor, position sensor, thermal sensor, speed sensor, acceleration sensor, energy consumption sensor, radar sensor, etc.

According to the working principle of the sensors, they are divided into vibration sensor, gas sensor, magnetic sensor, humidity sensor, vacuum sensor, etc.

According to the types of sensors output signal, they can be divided into analog sensor, digital sensor, switch sensor, etc.

According to the manufacturing process of sensors, they are divided into integrated sensors, thick film sensors, thin film sensors, ceramic sensors, etc.

According to the measurement target types of sensors, they are divided into chemical sensors, physical sensors, biological sensors, etc.

3.   Basic characteristics of sensors

Sensor characteristics refer to the corresponding relationship between the input and output of the sensor. Generally, sensor characteristics are divided into static characteristics and dynamic characteristics. The corresponding relationship between input and output of the sensor can be expressed by differential equations. The static characteristic of the sensor can be regarded as a special case of dynamic characteristic. In theory, when the differential terms of the first order and above in the differential equations are zero, the static characteristic can be obtained.

1)   Static characteristics

The static characteristics of the sensor refer to the relationship between output and input of sensor when its input signal is static. Because when the input signal is a static signal, the input and output of the sensor are not related to time. The algebraic equation without time variable can be used to express the static characteristics of the sensor, or the characteristic curve with the input and output as the coordinate axis can be used to express the static characteristics of the sensor. The main parameters representing the static characteristics of the sensor are as follows.

(1)   Linearity refers to the degree that the actual characteristic curve between the input and output of the sensor deviates from the fitting line. Its value is the ratio of the maximum deviation between the actual relationship curve and the fitting line within the range and the full range output value.

(2)   Sensitivity is expressed by the ratio of the increment of sensor output to the corresponding increment of input that causes the increment.

(3)   Hysteresis refers to the phenomenon that the input and output characteristic curves of the sensor do not coincide during the positive and negative stroke changes of the input. When the value of the input signal is the same, the value of the output signal of the positive and negative stroke of the sensor is not equal. The difference is the hysteresis difference.

(4)   Repeatability refers to the degree of inconsistency in the characteristic curve when the input of the sensor changes continuously in the whole range in the same direction.

(5)   Drift refers to the phenomenon that the output of sensor changes with time when the input is constant. The drift may be caused by the change of intrinsic parameters or environment of the sensor.

(6)   Resolution is the smallest increment of input signal that a sensor can detect. A certain input value changes gradually from a non-zero value. Only when the change of the input value exceeds the resolution, the output of the sensor will change. The sensor can distinguish the change of the input value at this time. Otherwise, the output of the sensor will not change. It cannot distinguish the change of the input value at this time.

(7)   Threshold, also known as threshold sensitivity, refers to the threshold that the output of the sensor does not change when the input is small enough. The threshold represents the signal resolution near the input zero.

(8)   Stability refers to the ability of a sensor to keep its performance parameters unchanged for a long time.

2)   Dynamic characteristics of sensors

The dynamic characteristics of the sensor refers to the corresponding relationship between the input and output when the input signal changes dynamically with time. The response of the sensor to the standard input signal can be easily measured by experiment. There is a specific relationship between the response and the response of the sensor and the non-standard input signal. The dynamic characteristics of the sensor are usually described by the response of the sensor and the standard input signal. The commonly used standard input signals are step signal and sine signal. Therefore, the dynamic characteristics of the sensor are mainly expressed as step response and frequency response.

2.4.2   Requirements of Sensors for IOT

The main·difference between IOT sensor and traditional sensor is that IOT sensor has certain intelligence, which can carry out simple signal processing and transmission. The sensor of IOT collects data, and the data is transmitted to the network center through the transmission medium connected with the sensor to complete the signal processing.

The sensor of IOT is generally called intelligent sensor. Intelligent sensor is a kind of sensor with microprocessor and can process information. Intelligent sensor has the function of collecting, processing and exchanging information. Intelligent sensor can use software technology to achieve high accuracy of information collection, achieve part of programming automation, low cost and diverse functions.

Because the IOT is limited by the external environment, intelligent sensors need to meet the conditions of miniaturization, low cost, low power consumption, anti-interference and so on.

① The characteristics of the IOT require that the sensor should be miniaturized with micron or nanometer size, gram or milligram mass, and cubic millimeter volume. ② The precondition of large-scale IOT is to reduce the production and operation costs, therefore, the cost reduction design method should be considered in the design of sensors to improve the production efficiency of sensors. ③ The sensor must adopt the power supply mode with low power loss to save energy. This is because the IOT takes the battery as its long-term energy source, which can use solar energy, light energy, biological energy to provide energy for the sensor. ④ Intelligent sensors need to be able to resist electromagnetic radiation, lightning, strong magnetic field, high humidity, obstacles and other adverse environmental interference. ⑤ Sensor nodes in the IOT can flexibly program applications according to software and hardware standards.

2.4.3 Sensor Network

1. Structure of sensor network

Sensor network is composed of multiple wireless sensor nodes, whose functions can be consistent or different. In the same area, many wireless sensor network nodes interact with each other by wireless way. The node can achieve data acquisition, data transfer and other functions in the network.

Sensor networks are often composed of sensor nodes, sink nodes and manager nodes. Sensor network uses sensor node group to supervise, manage and detect data, and different nodes transmit data in ad hoc way. In this process, the same data can be processed by multiple nodes. After multi hop transmission, the data will be collected to the sink node, and finally arrive at the management node through various communication networks. The end user can manage the data through the management node, and the user can also configure the sensor network and issue tasks through the management node. Figure 2.16 shows the system structure of wireless sensor network.

A large number of sensor nodes are randomly distributed in the monitoring area, and these sensor nodes construct the network in the form of self-organization. Sensor nodes generally use battery as energy of micro embedded system. Its data processing, storage, communication ability is weak. Each sensor node not only has the terminal information sending function, but also can forward the message. It completes the

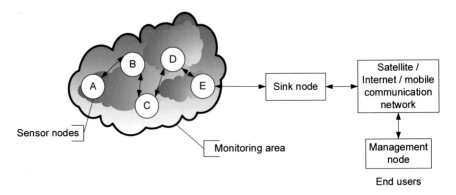

Fig. 2.16 System structure of wireless sensor network

local information collection and processing, as well as to store, manage, fuse the information transmitted by other nodes. The task needs to be completed by multiple sensor nodes cooperating with each other.

The capacity of information processing, storage and transmission of sink node is better than that of ordinary node. The sensor network is connected with the Internet and other external networks through the sink node. The information interaction between the management node and the network is completed through the communication protocol conversion between the two protocol stacks. The monitoring task submitted by the management node is released, and the collected data is transmitted to the external network. Sink node can provide enough energy, memory and resources, not only as a sensor node with enhanced function, but also as a special gateway device with only wireless communication interface.

The end users use the management node to manage and configure the sensor network, release the monitoring tasks to the sink node, and receive the monitoring data from the sink node.

2. Sensor node structure

The basic unit of wireless sensor network is sensor node. Sensor node is an important part of sensing, data processing and wireless communication. The node is powered by power module. The progress of sensor node technology is closely related to the development of wireless sensor network. Sensor node directly determines the performance of the whole network.

Sensor nodes include sensor module (such as sensor, A/D converter), data processing module (such as processor, memory), communication module (such as network, MAC, transceiver) and energy supply module (such as battery, natural energy), etc. Figure 2.17 shows the node structure of the sensor.

The sensor module senses and obtains the physical world information through the sensor, and converts the collected analog signal into digital signal through A/D converter, and then transmits the digital signal to the data processing module. The data processing module is responsible for adjusting and controlling each part of each

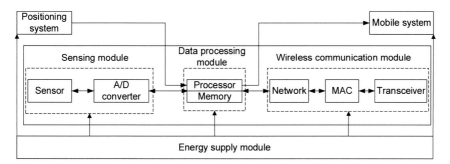

Fig. 2.17 Node structure of sensor

node, such as controlling the equipment, assigning tasks, achieving information transmission and fusion, etc. The processor and memory process and store the collected information. The communication module can realize the information interaction with other sensors or receivers. The energy supply module provides the necessary energy for the sensor to work normally. For the wireless network, it cannot use the ordinary electric energy, but can only use its own stored electric energy or the natural energy obtained from the outside.

2.4.4 Sensor Network Technology

The premise of the use of sensor network technology is that the network energy supply is limited, and its node equipment can monitor and control the target for a long time. Therefore, the energy efficiency of network function is the improvement direction of sensor network technology.

1.   Network topology control

For wireless and self-organizing sensor networks, network topology control is one of the key technologies. Using topology control, sensor networks can build a good network topology, which improves the efficiency of routing protocol and multiple access control protocol. Topology structure is conducive to prolong the network lifetime by saving node energy, and lay a good structural foundation for information fusion, time synchronization, target location and other applications.

The goal of wireless sensor network topology control is to meet the requirements of network connectivity and coverage. By controlling the power of the network, selecting the backbone nodes and their related mechanisms, and eliminating the useless wireless communication links between nodes, we can finally build a network topology that can transmit data efficiently.

Network node power control is to adjust the transmission power of nodes in the network to reduce the transmission power and balance the number of neighbor nodes that can be reached by a single hop under the premise of meeting the requirements of network connectivity. Wireless sensor network can adopt hierarchical topology control, use clustering mechanism, select some nodes as cluster heads, and then combine these cluster heads to form a backbone network that can process and forward data. The backbone node is the module node in the communication, while other non-backbone nodes can sleep temporarily and shut down their communication module to save energy.

The sleep/wake-up mechanism of the backbone node makes the node go to sleep when no communication is needed, and temporarily shut down the communication module. When communication is needed, it can wake up automatically in time and wake up the neighboring nodes at the same time. This topology control mechanism is mainly to solve the problem of state transition between dormant and active nodes, but it cannot be used alone and must be combined with other topology control.

2.   Network protocol

Network resources change with the dynamic changes of the topology of sensor networks. Each independent node in the network uses the sensor network protocol to form a multi-source data transmission network. The routing of sensor networks is data centric, and the network focuses on transmitting the data of all nodes to the sink node for processing, rather than interacting with a specific node. The function of network layer routing protocol is to transmit data packets from the source node to the destination node through the network. While considering the energy consumption of a single node, routing protocol also needs to evenly distribute the energy consumption to each node in the network, so as to extend the life cycle of the whole network.

The research focus of traditional wireless communication network is to improve the quality of service of wireless communication. Because the wire sensor nodes are randomly distributed and powered by batteries, the research focus of wireless sensor network routing protocol is to improve energy efficiency. Common routing protocols for wireless sensor networks include Flooding protocol, Gossiping protocol, SPIN protocol, directed diffusion protocol, LEACH protocol and so on.

1)   Flooding protocol

Flooding protocol is one of the initial protocols of wireless communication routing. According to the flooding protocol, each node in the network can receive information from other nodes, and send its own information to the neighbor nodes in the form of broadcast, and finally send the information data to the target node. The advantage of this protocol is that it is easy to implement and does not need to provide special information resources for routing control algorithm and network topology information. It is suitable for the network with high robustness. However, flooding protocol is prone to "information Implosion" and "information overlap" problems, resulting in information waste. Therefore, based on the flooding protocol, an improved protocol, namely visiting protocol, is proposed.

2)   Gossiping protocol

The Gossiping protocol improves the flooding protocol. It transmits information to a randomly selected neighbor node. The neighbor node that obtains the information selects another neighbor node arbitrarily by the same means to interact with it. In this way, the network information dissemination is completed. This kind of point-to-point information transmission avoids the energy consumption of transmitting information in the form of broadcast, but at the same time leads to the extension of information transmission time. Although Gossiping protocol alleviates the problem of network information implosion in flooding protocol to a certain extent, the phenomenon of network information overlapping still exists.

3.   SPIN protocol

In order to deal with the problems of information implosion and information overlap in flooding protocol and Gossiping protocol, an improved protocol spin appears. SPIN protocol is a kind of information dissemination protocol, which is based on node negotiation and can achieve energy adaptation. There are three types of spin messages: ADC, REQ and DATA.

(1)   ADC is used to broadcast data. When sharing the data information of a node, the data of this node is broadcast to other nodes through ADC.
(2)   REQ is used to request to send data. When a non-sensor node expects to get external data information, it sends REQ to request data.
(3)   DATA is a packet collected by a sensor. In the process of data packet transmission, firstly, the sensor node broadcasts data through ADC. If any node needs to receive the data packet, it sends REQ to the sensor node that broadcasts the data to request to receive the data, so as to establish the information interconnection between the sending node and the receiving node.

4.   Directed diffusion agreement

Directed diffusion protocol is also a kind of information dissemination protocol, the core of which is data. Directed diffusion algorithm can be divided into three stages, namely interest diffusion, gradient establishment and path enhancement. In the interest diffusion phase, sensor nodes receive information about the type and content of interest message, which includes task type, target area, data transmission rate, timestamp and other parameters. The sensor node stores the received interest messages in the cache. After all the interest messages are saved, a gradient field is constructed between the sensor node and the sink node, which is based on the principle of cost minimization and energy adaptation. Gradient field can optimize the path and achieve the fast transmission of data information.

5.   LEACH Protocol

LEACH is a layered protocol. Its goal is to balance the energy consumption of nodes in wireless sensor networks by randomly selecting class head nodes and evenly distributing the relay communication services of wireless sensor networks, so as to minimize the energy consumption of sensor networks and prolong the network lifetime. LEACH protocol can extend the network lifetime by 15%. LEACH protocol is divided into two processes: class preparation and data transmission. The total duration of these two stages is one whole cycle.

6. Network security

Wireless sensor network uses wireless transmission channel, like other wireless networks, also need to study its network security. There are many problems in sensor networks, such as eavesdropping, malicious routing, message tampering and so on. In the security design, considering the characteristics of sensor networks, we must pay attention to the following problems.

(1)   Due to the limited computing power and storage space of wireless sensor network, it is not suitable for the security algorithm with long key, high time and space complexity. We can use customized stream encryption and block encryption RC4/6 and other algorithms to design the security algorithm.
(2)   Before the layout of wireless sensor network, the connection between nodes is unknown, and the knowledge base of later node layout is lacking. Therefore, the public key security system is not suitable for wireless sensor network. It is difficult to realize the point-to-point dynamic security connection.
(3)   When formulating the security mechanism of wireless sensor networks, it is necessary to consider the related security issues of all networks.

7. Synchronous management

Synchronization management mainly refers to clock synchronization management. In simultaneous interpreting, the sensor nodes in wireless sensor networks are independent of each other, and the frequency of crystal oscillations between different sensor nodes is different. The running time of nodes will also be affected by the external environment temperature and electromagnetic waves. Wireless sensor network (WSN) is a distributed network system, which needs the cooperation among the network nodes to complete the task. Therefore, to ensure the time synchronization of each node is an important part of the synchronization management mechanism.

In order to synchronize the local clocks of all sensor nodes in wireless sensor networks, the energy efficiency, scalability, robustness, accuracy, effective synchronization range, cost and size of the network need to be considered when formulating the time synchronization mechanism.

8. Positioning technology

The location information is an important message that sensor nodes get when they collect data. The controlled message is valid only if it has the location information. Therefore, one of the basic functions of sensor networks is to locate the target information. In order to make the location information of randomly distributed sensor nodes accurate, they must be able to complete their own localization after deployment. Because sensor nodes have the characteristics of limited resources, random distribution and weak anti-interference ability, the localization principle must meet the requirements of self-organization, robustness and distributed computing.

Sensor nodes are divided into beacon nodes and unknown location nodes according to whether they know the location of their nodes. Beacon node is a node with known location, and unknown location node should determine its own location according to some beacon nodes with known location and specific positioning methods, such as trilateration, triangulation, maximum likelihood estimation, etc. According to whether the distance or angle between nodes is measured in the positioning process, the positioning in sensor networks is classified as distance based positioning and distance independent positioning.

Distance based positioning mechanism is to determine the location of unknown nodes by measuring the actual distance between adjacent nodes. It often realizes the positioning function by measuring the distance, determining the orientation and correcting the position. Distance based localization mechanisms can be classified according to the different application methods when measuring the distance between nodes. The applied localization methods include TOA (time of arrival), TDOA (time difference of arrival), AOA (angle of arrival), RSSI (received signal strength indication), etc. The distance based positioning mechanism needs to measure the distance between nodes. It requires higher positioning accuracy and higher hardware requirements.

Distance independent localization mechanism can determine the orientation of unknown nodes without measuring the absolute distance or orientation between nodes. Distance independent localization mechanism mainly includes centroid algorithm, DV-Hop algorithm, Amorphous algorithm, APIT algorithm and so on. Because there is no need to measure the absolute distance or orientation between nodes, the requirement of node hardware is relatively low. It is more suitable for large-scale sensor networks. The distance independent positioning mechanism is less affected by environmental factors. Although the positioning error will increase correspondingly, its positioning accuracy can still meet the requirements of most sensor network applications.

9.   Data fusion

In order to effectively save energy and reduce the amount of data transmitted in sensor networks, in the process of data acquisition of each sensor node, the local computing capacity and storage capacity of each node can be integrated to remove redundant information and achieve energy saving. In order to solve the problem that sensor nodes are prone to failure, sensor networks need to integrate information data and process them comprehensively to improve the accuracy of information. Data fusion technology can save energy, improve the efficiency of data collection and obtain accurate information.

Data fusion technology can be used in multiple protocol layers of sensor networks. In the network layer of sensor networks, most routing protocols use data fusion mechanism to reduce the amount of data transmitted, use distributed database technology to plan the network application layer, gradually screen the collected data, and then complete the data integration.

Data fusion technology can greatly improve the accuracy of information, and has a good effect in saving energy consumption. But it is followed by the growth of data transmission time and the reduction of network robustness. In the process of data transmission, variety ways can be used to increase the network delay, including searching for the route responsible for data fusion, waiting to receive the data to be fused, and carrying out data fusion related operations. Compared with the traditional network, sensor network nodes are easy to fail, and their data is easy to lose. Data redundancy can be increased through data fusion, but the rise of data loss rate will also lead to the loss of effective information, and the robustness of the network will be reduced.

10. Data management

From the perspective of data storage, sensor network is a kind of distributed database. The sensor network manages its information database and separates the logical view stored in the sensor network from the practical application of the network. Therefore, users only need to pay attention to the logical structure of data query and do not need to study the detailed operation of how to realize the network. Abstracting network data will reduce the efficiency of data processing to some extent, but greatly increase the usability of sensor networks. Typical sensor data management systems are represented by tiny DB system of UC Berkeley and Cougar system of Cornell University.

Although sensor network uses database to manage data, it is different from traditional distributed database. Because sensor nodes are vulnerable to failure and have limited energy, the data management system of sensor networks must reduce the energy loss when providing effective data services. There are two reasons why the traditional distributed database management system cannot be applied to the sensor network. On the one hand, there will be unlimited data flow on the sensor network nodes, and the number of nodes is particularly large. On the other hand, continuous query method or random sampling query method are usually used in data query. The structure of data management system in sensor network includes centralized, semi-distributed, distributed and hierarchical, among which there are many researches on semi distributed structure.

In sensor networks, data storage is achieved by external storage, local storage and data centric storage. Compared with the other two methods, the data centric storage method has better communication efficiency and lower energy consumption. Data centric data storage is a typical data storage method based on geographic hash table.

It can build dimensional index for data query in sensor network. DIFS (distributed index for features in sensor network) system data index dimension N is equal to 1. DIM (distributed index for multidimensional dam) is a multidimensional index method. Generally, SQL like language is used as the data query language of sensor network. There are many kinds of query methods in sensor networks. Centralized query can transmit a large amount of data, including redundant data. Its energy consumption is high. Distributed query and pipeline query both use aggregation technology, and the energy consumption of data transmission of distributed query is low, while pipeline query has better aggregation accuracy than distributed query.

11. Wireless communication technology

Wireless communication in sensor networks requires low power consumption and short distance. IEEE 802.15.4 standard is adopted as the wireless communication standard for low-speed wireless personal area network. IEEE 802.15.4 standard provides a unified standard for low-speed networking, which is suitable for the networking needs of different devices within the scope of individuals or families. As a kind of wireless communication technology, ultra wideband (UWB) technology has great potential. UWB technology has many advantages, the channel attenuation has little effect on it. The UWB signal transmission power spectrum density is low, and the system structure is relatively simple. Information transmission process is not easy to be intercepted, thus, the positioning accuracy can reach centimeter level. Therefore, UWB technology is suitable for wireless sensor networks.

12. Embedded operating system

The sensor node in sensor network is a micro embedded system. When using the memory, CPU and communication module of the system, due to the limited hardware resources, the system not only needs to be energy-saving and efficient, but also needs to be able to fully support specific applications. Each specific application is based on the embedded operating system of wireless sensor network, which works in the same period of time and makes use of the limited information resources of the network.

Sensor nodes have two obvious advantages: high concurrency and high modularity. In sensor networks, it is possible to carry out multiple logic controls at the same time. These logic controls are usually carried out frequently. The execution process is short, and the concurrency density of sensor nodes is high. These requirements of sensor nodes should be fully considered when designing the operating system. Most sensor nodes are integrated into modules for combined application, thus, the operating system needs to control the hardware conveniently. In addition, the application should have the function to facilitate the reconstruction of each part without additional consumption to the system.

13. Application layer technology

The application layer of sensor network is composed of many kinds of software systems, and the set sensor network can usually complete many tasks.

Through the integration of a variety of sensors and low-speed, short-range wireless communication technology, a special network is constructed, which is composed of many sensor nodes. These sensor nodes have low power consumption and small size. They can communicate by wire as well as wirelessly. At the same time, they also have good computing power. Sensor networks can achieve the information interaction between things in a specific area, and provide technical support for the intelligent perception, data acquisition and other functions of the IOT technology. It is an important part of the IOT.

2.5   IOT and Internet

The IOT is a special form of Internet with Internet as the center and root, which is the extension and expansion of the Internet. The Internet is an essential part of the IOT. Without the Internet, there will be no IOT.

2.5.1   Internet

Internet is an information system based on TCP/IP protocol to build a unique address logic to cover the world. It is an international computer network constructed by WAN, LAN and single computer according to a certain communication protocol. It can realize not only the transmission control protocol, but also the transmission internet protocol. As an international computer network connected by multiple computer networks according to specific protocols, the Internet can achieve the interconnection of network resources without geographical limitations, and improve the convenience of information exchange. The IOT is not an actual network body with special network boundaries. It is generally believed that the network set connected by gateway is an IOT. The computer network of the Internet is composed of LAN, MAN and WAN. Through various communication lines, such as ordinary telephone line, high-speed dedicated line, satellite, microwave and optical cable, the network resources of different regions, industries, institutions and individuals are connected and combined to achieve information exchange and transmission and resource sharing.

As an electronic identifier, the IOT is mainly composed of software and chip. Through the software and embedded chip, the IOT has the ability of execution, sensing and computing. The IOT uses sensor devices to obtain information and data of related items. The IOT can connect the Internet and things at any time and place, and transmit resource information between things and people, things and things, so as to achieve the management and control of things.

The Internet was established by the United States in 1969 for military connection. Because of its universal applicability, it is used in schools and scientific research institutions. With the development of computer technology, the Internet began to be used in business. It has great potential in data communication, information retrieval, user service and so on, and its application scope is more and more extensive.

Under the unified TCP/IP protocol, the network connected by computer is called Internet 1.0 era. In the 1.0 era, the Internet takes the computer as the main node. With the birth of the World Wide Web hypertext transfer protocol, the richness and friendliness of the interface are improved. The retrieval function is more convenient and efficient. More and more people use the Internet, especially the progress of communication technology, which makes the Internet develop rapidly.

With the development of Internet 2.0 era, it is no longer the main purpose of the Internet to connect computers. At this time, the Internet aims at serving people, and the main node of the Internet is people. In the era of Internet 2.0, the shared content is not limited to computer resources. The development of Internet promotes the economic development and policy reform. However, the Internet has the shortcomings of information symmetry, timeliness, reliability and sharing, which need to be improved.

The IOT is developed on the basis of the Internet. In 1999, scientists from Massachusetts Institute of Technology used RFID technology to achieve object to object communication, and proposed that all things can be connected by network. With the development of computer technology, communication technology and sensor technology, the ways of connecting things are becoming more and more diversified. People can not only use RFID devices, but also use other information sensing devices, such as QR code reading devices, infrared sensors, laser scanners, etc., to achieve the interaction and communication between things according to a certain protocol, and achieve the functions of intelligent identification, positioning, tracking, monitoring and management of the IOT. The IOT is a network of perception, connection and interaction. The network form characterized by the IOT is the 3.0 era of the Internet.

2.5.2 The Relationship Between Internet and IOT

The IOT is based on the Internet, through the induction equipment to identify the external objects and obtain data resources, using a certain standard to connect things with the Internet, thus forming a way of communication. The development of Internet technology is relatively mature, and its information transmission is efficient and reliable. At this stage, more mature Internet information and communication technology in China is gradually developing towards the direction of the IOT, realizing the connection between things and things, people and things, which greatly facilitates life and work.

The IOT and the Internet are two programs running on the same platform, both of which rely on network technology for information and data transmission, and can independently complete tasks in the process of operation. The Internet platform pays more attention to the transmission of information and data. They are relatively equal. The IOT does not attach great importance to the dissemination of information. It pays more attention to the content of efficiency and management. The IOT system contains a number of complementary and related tasks. Different from the Internet, the IOT pays more attention to the actual relationship between people and things or things and things. Information security of IOT is stronger. The IOT is an extension of the traditional Internet network technology, which integrates virtual and real things. Its purpose is to use the network to help people realize the control of things in real life.

According to the different stages of IOT application, the relationship between IOT and Internet can be divided into three types: independent, cross and inclusive.

(1) IOT and Internet are independent of each other. Some experts believe that the IOT perceives objects in the external environment through sensors and forms an interconnected sensor network, but this sensor network is independent of the Internet. For example, the sensor network of Shanghai Pudong Airport, which is independent of the Internet, is known as the first IOT in China. From this point of view, the IOT and the Internet are two independent networks.

(2) The IOT and the Internet intersect, and the IOT is the complementary network of the Internet. The main body of Internet is different from that of IOT. Internet is a global network formed by computer connection, which can achieve the information transmission between people, and its main body is people. The main body of the IOT is a variety of things, through the network to transmit information between things, and ultimately achieve the purpose of serving people. The IOT is the expansion and supplement of the Internet.

(3) The Internet includes the IOT. In this relationship, the Internet is not the basic form of the present, but a form of the future. With the continuous development and progress of the Internet, its future form is a network form containing all things, and the IOT is subordinate to this network form.

2.5.3 Comparison of Basic Characteristics Between IOT and Internet

1. The similarity of network performance between Internet and IOT

(1) The technology of Internet and IOT rely on the same network technology foundation.

(2) Internet and IOT work on the basis of the same packet data technology.

(3) Internet and IOT have a common bearer network, and their bearer network and business network are separated from each other, business network can be expanded and designed independently.

2. The difference of network performance between Internet and IOT

(1) As a traditional network communication technology, the Internet focuses on the accessibility and exploitability of the network, as well as its own resource management and transmission capabilities, while the requirements for network efficiency and management quality are not clear enough.

(2) Compared with the Internet, the IOT has higher requirements for the network. There are many independent subsystems in the IOT system. In the working process of these subsystems, the network is required to have high real-time, reliability, security and effectiveness.

(3) The IOT and the Internet have different network organization forms. The IOT connects the information of things and realizes the communication between things. At the same time, IOT technology can control and manage things according to the different network connection methods.

3.   Application comparison between Internet and IOT

As an indispensable part of social development at this stage in China, the Internet has certain virtuality. In the process of practical application, additional devices need to be added to improve the value of enterprise network applications. The Internet has the characteristics of large-scale and high consumption, which leads to the high cost of enterprise network construction and operation.

The IOT is the operation of the actual existence of things. In the process of practical application of the IOT, it is necessary to install the corresponding electronic identifier (such as chip) on each measured object to achieve better network control effect, which leads to higher application cost of the IOT, and different from the Internet. It is difficult to reduce the application cost of the IOT in a short time. Due to the limitations of the methods and technologies of the IOT, its products and services are relatively poor compared with the Internet.

4.   Comparison of technical standards between Internet and IOT

There are RFID terminals and sensor networks in the sensing layer of the IOT. Its network structure is more complex than that of the Internet. The key technologies of the IOT are diverse and widely used. The standard system of the IOT is relatively complex, while the technical standard of the Internet is relatively single. Different from the IOT, the Internet has a unified terminal, and its terminal service is undifferentiated service, which will not put forward higher requirements for servers and computers. Although the Internet lacks a unified control center, its degree of autonomy is higher than that of the IOT.

Chapter 3
Information Security and Big Data

3.1 Encryption Algorithm

3.1.1 Basic Concepts of Safety

1. The connotation of information security

In modern times, the connotation of information security after the emergence of the Internet includes data oriented security, that is, the protection of information confidentiality, availability and integrity. As well as user oriented security, that is, authentication, authorization, access control, anti-repudiation and serviceability. And also the protection of personal privacy and intellectual property rights.

Information security is very important in the IOT system, which is related to whether the information, equipment and decision-making system of the whole IOT system can run safely and reliably.

2. The principle of safety

Security principles include confidentiality, authentication, integrity, non-repudiation, access control and availability. Among them, the principle of confidentiality requires that only the sender and the receiver can access the information content. The principle of authentication requires that the source of the document and the electronic information be correctly identified. We can establish the identity certificate through the authentication mechanism. The lack of authentication mechanism may lead to attacks. The principle of integrity requires that the information content cannot be sent from the sending to the receiving. In this process, if the user is attacked by modification, the integrity of the information will be lost. The non-repudiation principle requires that the user cannot deny the information after sending, that is, the sender is not allowed to refuse to acknowledge the information sent. The access control principle can specify what the control user can access. The availability principle can specify the authorized party to provide information at any time, and the interruption attack will destroy the availability.

© Publishing House of Electronics Industry 2022
B. Wang et al., *Internet of Things and BDS Application*,
https://doi.org/10.1007/978-981-16-9194-2_3

3.   Types of attack

The types of attack can be divided into two types, active attack and passive attack. Passive attack refers to the attacker eavesdropping or monitoring the transmission of data in information transmission. This attack will not modify the data. Figure 3.1 shows the classification of passive attacks. Active attack is to modify information content or generate false information in some way. This kind of attack can be found and recovered, but it is difficult to prevent it. Active attack is divided into camouflage attack, modification attack and forgery attack. The classification of active attack is shown in Fig. 3.2.

In practice, attack types can be divided into application layer attack and network layer attack. It may occur in both application layer and network layer. It can be divided into viruses, worms, Trojans, small programs and ActiveX controls, cookies, JavaScript, etc. All of these may cause attacks on computer systems.

Fig. 3.1 Classification of passive attack

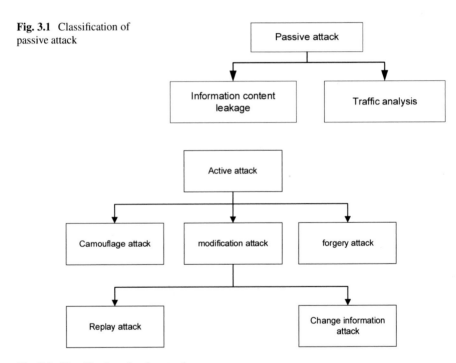

Fig. 3.2 Classification of active attack

3.1.2 Encryption Algorithm

When the IOT system transmits information, especially when it transmits wireless information, it needs to consider the situation that information is intercepted. Therefore, for the information of some important facilities and equipment, because of its sensitivity, the transmitted information must be encrypted. We first introduce the basic concepts of cryptography. The mechanism of encoding information for safe transmission is called cryptography. Cryptography encodes messages unreadable, so as to ensure the security of messages.

1. Plaintext and ciphertext

Plaintext is the information that can be understood by sender, receiver and anyone who accesses the information, while ciphertext is obtained by encoding plaintext message. The process of obtaining plaintext message from ciphertext message is called cryptanalysis. There are usually two methods to convert plaintext message into ciphertext message: substitution method and transformation encryption technology.

2. Substitution method

Substitution method is to replace the character of plaintext message with another character, number or symbol during encryption. It can be replaced by single code encryption method, homophone replacement encryption method, block replacement encryption method and multi code replacement encryption method. Among them, the single code encryption method has a fixed replacement mode. That is, every letter in the plaintext is replaced by the letter in the ciphertext. Mathematically, it can use any replacement and combination of 26 letters, that is, there are 4×10^{26} possibilities. Due to the large amount of replacement and combination, the single code encryption method is difficult to crack. Homophonic Substitution Cipher is to replace a letter with another letter, but a plaintext letter in homophone substitution encryption may correspond to multiple ciphertext letters, such as Z replaced as A, B, E, F, G, etc. Polygram substitution cipher is to replace a block of letters of plaintext with another block of letters. Polyalphabetic Substitution Cipher uses multiple single code keys, and each key encrypts a plaintext character. After all the keys are used up, it will continue to be recycled. If there are 10 single code keys, every 10 letters in the plaintext will be replaced with the same key, and 10 represents the ciphertext period.

3. Transform encryption technology

Transformation encryption technology can be used to replace plaintext letters, such as rail fence encryption technology, simple columnar transformation encryption technology, Verman encryption method and running key encryption method.

The algorithm of Rail Fence Technique can be divided into two steps. The first step is to write the plaintext message into diagonal sequence. The second step is to read the plaintext written in the previous step into line sequence. That is, to generate ciphertext line by line.

In the Simple Columnar Transposition Technique, the algorithm can be divided into two steps. The first step is to write the plaintext message line by line into a rectangle of predetermined length. The second step is to read the information column by column in random order, and the final message is the ciphertext message. On this basis, in order to make the password more difficult to decipher, we can transform the Simple Columnar Transposition Technique for many times to enhance its complexity.

Verman encryption method uses a random set of non-repetitive characters as the input ciphertext. Once the method uses the transformed input ciphertext, it will not be reused. It is also known as One-Time Pad. In this method, each plaintext letter is arranged as a number in increasing order. The input ciphertext letter is processed in the same way, then the plaintext and the corresponding letter in the ciphertext are added (if the sum is greater than 26, 26 should be subtracted from the plaintext itself and then transformed), and finally converted into the corresponding letter to get the output ciphertext.

Running Key Cipher uses a text character in the book as a one-time password when generating ciphertext, and adds it with the input plaintext message.

4.   Encryption and decryption

Encryption is the process of changing plaintext information into ciphertext information, while decryption is the opposite of encryption. That is, decryption is the process of changing ciphertext information into plaintext information. Figure 3.3 shows the encryption and decryption process.

In communication, the sender needs to encrypt the plaintext information, and then transmit it to the receiver through the network. The receiver will restore the encrypted ciphertext information to plaintext information. Encryption algorithm should be used to encrypt plaintext information. Similarly, decryption algorithm should be used to decrypt received encrypted information. Decryption algorithm should correspond to encryption algorithm, otherwise the original information cannot be obtained through decryption. The algorithms in the encryption and decryption process can be made public, but in order to ensure the security of the encryption process, the key is needed. According to the key used, there are two mechanisms: symmetric key encryption and asymmetric key encryption. The difference between the two is that symmetric key

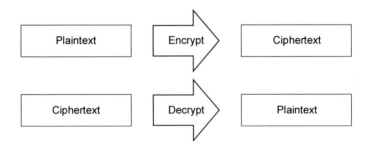

Fig. 3.3 Encryption and decryption process

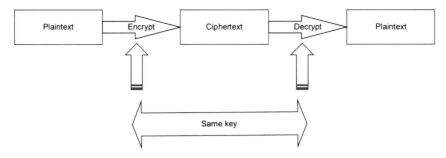

Fig. 3.4 Encryption process of symmetric key

encryption uses the same key for encryption and decryption, while asymmetric key encryption uses different keys for encryption and decryption.

5.   Symmetric and asymmetric key encryption

Symmetric key encryption is also called private key encryption or shared key encryption. The sender and receiver use the same key to lock and unlock the data, and password distribution is associated with symmetric key distribution. The encryption and decryption of asymmetric encryption algorithm use two different keys: public key and private key. The public key and the private key are a pair. If the public key is used to encrypt the data, only the matching private key can be used to decrypt. If the private key is used to encrypt the data, only the same matching public key can be used to decrypt. Because encryption and decryption use two different keys, this algorithm is called asymmetric encryption algorithm.

The basic process of the algorithm is as follows. Party A generates a pair of keys and publishes one of them to the outside as a public key. Party B who obtains the public key encrypts the confidential information with the key and then sends it to Party A. Party A decrypts the encrypted information with another private key saved by itself. Figure 3.4 shows the encryption process of symmetric key, and Fig. 3.5 shows the encryption process of asymmetric key.

3.1.3   Symmetric Key and Asymmetric Key Encryption Algorithm

The basic process of data encryption is to process the original plaintext file or data according to some algorithm, so that it becomes an unreadable code. Only after the corresponding key is input, the original content can be displayed. In this way, the purpose of protecting data from being stolen and read can be achieved. Encryption algorithm can be divided into symmetric key encryption algorithm and asymmetric key encryption algorithm. These encryption algorithms have two key aspects: algorithm type and algorithm mode.

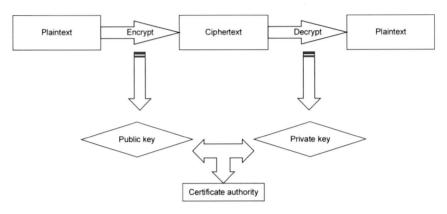

Fig. 3.5 Encryption process of asymmetric key

Algorithm type defines the length of plaintext in each step of encryption algorithm. There are two methods to generate ciphertext: stream ciphers and block ciphers. Stream encryption encrypts only one bit of plaintext at a time, and decrypts bit by bit when decrypting. Block encryption encrypts one block of plaintext at a time, and decrypts block by block when decrypting. In the algorithm, the word "group" denotes the number of changes when plaintext generates ciphertext. The concept of confusion in encryption algorithm is to ensure that there are no clues related to plaintext in ciphertext and prevent cryptanalysis from finding the corresponding plaintext from ciphertext. The concept of diffusion in encryption algorithm can increase the redundancy of plaintext. For example, stream encryption only uses obfuscation, while block encryption can use both obfuscation and diffusion.

The algorithm pattern in encryption algorithm defines the details of algorithm encryption in specific types. There are four algorithms: electronic code book (ECB), cipher block chaining (CBC), cipher feed back (CFB) and output feedback (OFB). Figure 3.6 shows the algorithm mode in the encryption algorithm.

In electronic codebook mode, the input plaintext information is divided into 64 bit blocks, and then each block is encrypted separately. All blocks in the message

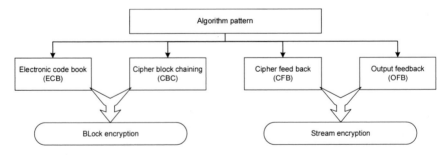

Fig. 3.6 Algorithm mode in encryption algorithm

are encrypted with the same key. The advantage of electronic codebook mode is simple, which is conducive to parallel computing, and the error will not be spread. Its disadvantage is that it cannot hide the plaintext mode, which may cause active attack on plaintext. Therefore, this mode is suitable for encrypting small messages.

Encrypted block link mode ensures that even if the plaintext blocks in the input are repeated, these plaintext blocks will get different ciphertext blocks in the output. The advantage of encrypted block link mode is that it is not vulnerable to active attack, and its security is better than that of ECB. It is suitable for transmitting long length messages. It is the standard of SSL and IPSEC. Its disadvantage is that it is not conducive to parallel computing, the error will be transferred and the vector N needs to be initialized. Not all applications can handle application blocks. Character applications also need security.

In encryption feedback mode, the data is encrypted in smaller units, which is less than the defined block length (usually 64 bits). The advantage of the encryption feedback mode is that it hides the plaintext mode, transforms the block cipher into the stream mode, and encrypts and transmits the data smaller than the block cipher in time. Its disadvantage is that it is not conducive to parallel computing, and the error transmission needs the unique N.

The difference between the output feedback mode and the encryption feedback mode is that the ciphertext is filled in the next stage of the encryption process in the encryption feedback mode. But in the output feedback mode, the output of the encryption process is filled in the next stage of the encryption process. The advantage of output feedback mode is that plaintext mode is hidden, block cipher is transformed into stream mode, and data smaller than block can be encrypted and transmitted in time. Its disadvantage is that it is not conducive to parallel computing, may cause active attacks on plaintext, and there is the possibility of error transmission (a plaintext cell is damaged, affecting multiple cells).

1.   Symmetric key encryption algorithm

In symmetric key encryption algorithm, the sender of data transforms plaintext and encryption key into complex encrypted ciphertext after special encryption algorithm. After receiving the ciphertext, the receiver decrypts the ciphertext with the key used in encryption and the inverse operation of the same algorithm, and recovers it to readable plaintext. In symmetric encryption algorithm, only one key is used. Both sender and receiver use this key to encrypt and decrypt data, which requires the decryptor to know the encryption key in advance. The advantages of symmetric encryption algorithm are small computation burden, fast encryption speed and high encryption efficiency. But the disadvantages are obvious. Because both sides of the transaction use the same key, the security cannot be guaranteed.

Data encryption standard (DES), also known as data encryption algorithm (DEA), is a very popular symmetric key encryption algorithm. DES usually uses ECB, CBC or CFB mode.

DES uses a 56 bit key, but its initial key number is 64 bits. Before the DES process starts, each 8th bit of the key is discarded to get a 56 bit key. Before giving up the key, in order to ensure that the key does not contain any errors, parity check can be performed on the number of bits to be given up.

The basic principle of data encryption standard is: DES is a block encryption method, which takes 64 bit plaintext as the input of DES and produces 64 bit cipher-text output. The two basic operations of DES encryption are substitution and trans-formation (also called confusion and diffusion). DES consists of 16 steps, each step is called a round. Replacement and transformation are proceeded in every round. The main steps of DES can be divided into the following six steps.

(1)   The 64 bit plaintext block is sent to the initial permutation (IP) function;
(2)   First, the plaintext is replaced;
(3)   The initial permutation generates two parts of the transform block, which are assumed to be left plaintext (LPT) and right plaintext (RPT);
(4)   Each LPT and RPT has its own key after 16 rounds of encryption;
(5)   Finally, the LPT and the RPT are reconnected, and their blocks are finally permutated (FP);
(6)   The 64 bit ciphertext is obtained.

The main steps of DES are shown in Fig. 3.7.

The initial replacement occurs only once, before the first round. After the comple-tion of the initial replacement, the 64 position text block is divided into two parts, each with 32 bits. The left block is called left plaintext (LPT) and the right block is called right plaintext (RPT). Then the two rounds are operated for 16 rounds. Figure 3.8 shows the steps of a round of DES.

Step 1: key transformation

Fig. 3.7 The main steps of DES

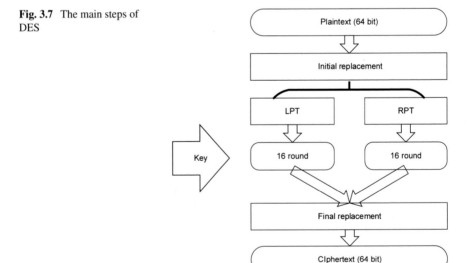

Fig. 3.8 The steps of a
round of DES

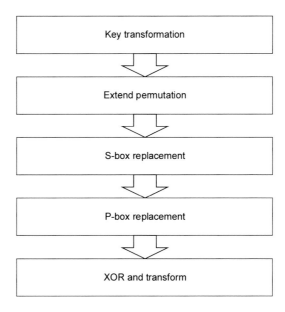

Each 8th bit is discarded by 64 bit key, so that there are 56 keys in each round. Key transformation is to make each round of 56 transforms produce different 48 bit subkeys. The specific method is to divide each round of 56 bit key into two parts. Each with 28 bits, and move one or two bits to the left. After shifting, 48 bits are selected from 56 bits. Because key transformation is to replace and select 48 bits of 56 bits, it is called compression permutation. In this process, different key bit subsets can be used in each round, which makes DES more difficult to decipher.

Step 2: extend permutation

After the above initial transformation, two 32-bit plaintext regions are obtained, which are called left plaintext and right plaintext respectively. The extended permutation not only extends the right plaintext from 32 to 48 bits, but also permutes these bits. The process is as follows.

First, divide the 32-bit right plaintext into 8 blocks, with 4 bits in each block, as shown in Fig. 3.9.

Then, each 4-bit block in the previous step is expanded into a 6-bit block. The extra two bits are the first and fourth bits of the previous 4-bit block (see Fig. 3.10). The second bit of the 4-bit block is written out as the password input of the 4-bit block. The second input bit and the 48th bit repeat the first input bit of the 4-bit block. That is, the first input bit of the 4-bit block appears in the second input bit and the 48th bit of the 6-bit block. The second input bit of the 4-bit block moves to the third input bit of the 6-bit block. This process is actually the expansion and replacement of the output bit when generating the output. Figure 3.10 shows the expansion and replacement process of the right plaintext.

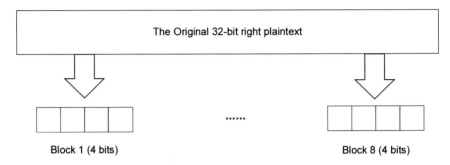

Fig. 3.9   The 32-bit right plaintext is divided into eight 4-bit blocks

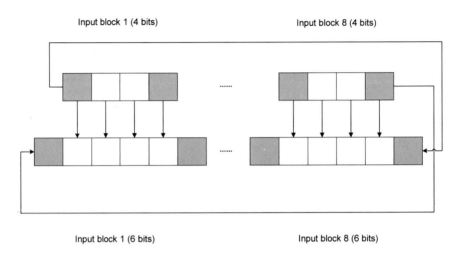

Fig. 3.10   Right plaintext extended permutation procedure

Step 3: S-box replacement

The process of S-box replacement is to get 32-bit output from 48 bit input obtained by XOR operation of compressed key and extended right plaintext. The replacement uses 8 boxes. Each S-box has 6-bit input and 4-bit output. The 48 bit input block is divided into 8 sub blocks (each with 6 bits). Each sub block specifies an S-box. The S-box turns 6-bit input into 4-bit output. Figure 3.11 shows the S-box replacement process.

The outputs of all S-boxes form a 32-bit block, which is transferred to the next stage, i.e. P-box replacement.

Step 4: P-box replacement

P-box replacement is to replace one bit with another according to the P table. It is a compression without expansion. This replacement mechanism only performs simple permutation.

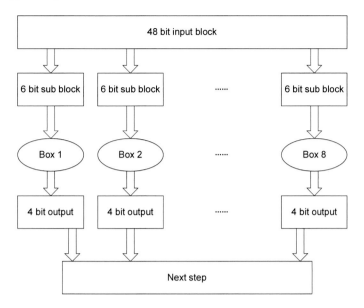

Fig. 3.11 S-box replacement process

Step 5: XOR and transform

The first four operations only deal with the right plaintext, but not the left plaintext. The XOR operation is performed on the left half of the original 64 bit plaintext and the result of P-box replacement to generate a new right plaintext. At the same time, the original right plaintext is changed into a new left plaintext in the process of exchange. Figure 3.12 shows the XOR and replacement process.

The next round shown in the module in Fig. 3.12 requires a total of 16 times, after which the final replacement will take place. The final permutation takes place only once, and its output is a 64 bit encrypted block. Because the values and operation order of each table in DES are carefully selected, the algorithm is reversible, and the encryption algorithm is also suitable for the corresponding decryption algorithm. In this process, it should be noted that the key part needs to be reversed. The strength of DES depends on its key. The algorithm uses 56 bit key, that is, it can generate $2^{56} = 7.2 \times 10^{16}$ keys.

According to the advantages of DES, some methods can be used to improve and deform DES, mainly including double DES and triple DES. Dual DES uses two keys K1 and K2. First, K1 is used to DES the original plaintext to get the encrypted text, and then K2 is used to DES the encrypted text again. In fact, the original plaintext is encrypted twice with different keys. Similarly, the decryption process is to decrypt twice in reverse order. In order to make the encryption method more powerful, triple DES appears after dual DES. Triple DES is divided into two categories, one uses three keys, the other uses two keys. The triple DES of three keys first uses key K1 to encrypt plaintext block, then uses key K2 to continue encryption, and finally uses key

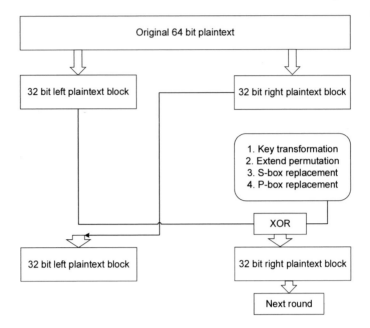

Fig. 3.12 XOR and substitution process

K3 to encrypt. However, the functions of keys K1, K2 and K3 are different. Triple DES with three keys is quite secure, but it also has a disadvantage. The number of keys is 168, which will cause great difficulties in practical application. Triple DES with only two keys proposes that the output of plaintext block is encrypted with key K1. Then the above output is decrypted with key K2, and then the decrypted output is encrypted with key K1 again. The purpose of decryption is to make triple DES use two keys instead of three keys. Unlike double DES using K1 and K2, triple DES with two keys will not be attacked by the middleman.

International data encryption algorithm (IDEA) is one of the most powerful encryption algorithms. IDEA is protected by patent and can only be used in commercial applications after obtaining a license. The famous e-mail privacy technology PGP is based on IDEA. Although IDEA is powerful, it is not as popular as DES.

RC5 symmetric block cipher algorithm (RC5 algorithm for short) is a block cipher algorithm with variable parameters. The three variable parameters of the algorithm are: block size, key size and encryption rounds. The advantage of RC5 algorithm is fast operation speed, only using basic computer operation. The number of rounds is variable, and the number of password bits is variable too, which greatly increases its flexibility to a certain extent. Applications with different security can flexibly set these values. The algorithm needs less memory, which is suitable not only for desktop computers, but also for smart cards and other devices with small memory. RC5 algorithm is a relatively new algorithm. Its developers have designed a special implementation of RC5 algorithm. Therefore, RC5 algorithm has a word oriented structure: RC5 $- w/r/b$, here w is the word length, and its value can be 16, 32 or 64. For different word lengths, the length of plaintext and ciphertext blocks is $2w$ bits, r is the number of encryption rounds and b is the length of key bytes. In RC5 algorithm, $2r + 2$ key related 32-bit words are used. The process of creating the key group is very complicated. First, we copy the key byte into the 32-bit word array, and if necessary, fill the last word with zero. Then we initialize the array, and encrypt the plaintext after creating the key group. When encrypting, first we divide the plaintext into two 32-bit characters A and B, and output the ciphertext in registers A and B. When decrypting, the ciphertext is divided into two characters A and B (the storage method is the same as encryption).

The operation principle of RC5 algorithm is complex. In the one-time initial operation, the input plaintext block is divided into two 32-bit blocks A and B. The first two sub keys S[0] and S[1] are added to A and B respectively to generate C and D, which indicates the end of one-time operation. Then we start the next rounds, and complete the shift and left shift in each round in turn. After that, we add the next subkey to C and D (display the addition operation, and then modulus the result with 2^{32}). The initial operation of RC5 algorithm includes the following steps: the input plaintext is divided into two equal length blocks A and B, and then the first sub key S[0] is added to A, and the second sub key S[1] is added to B. These operations are modulated with 2^{32} to obtain C and D respectively. Since each round of RC5 algorithm is the same as the first round, only the details of the first round are introduced here.

The first round of algorithm details can be divided into the following seven steps.

Step 1: C XOR D to get E (see Fig. 3.13)

Step 2: cycle shift E to the left for D bits (see Fig. 3.14)

Step 3: add E and the next subkey. The first round is S[2], and the ith round is S[2i]. The result of this operation is F (note that the operations from step 4 to step 6, whether XOR, cyclic left shift or addition, are the same as those from step 1 to step 3, only the input is different) (see Fig. 3.15)

Fig. 3.13 First step of each round

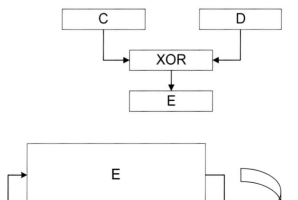

Fig. 3.14 The second step of each round

Fig. 3.15 The third step of
each round

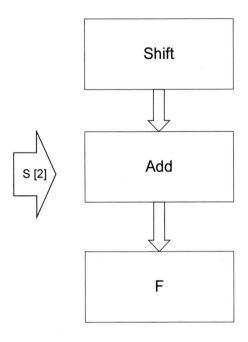

Step 4: D XOR F to get G (see Fig. 3.16)

Step 5: cycle shift G to the left for F bits (see Fig. 3.17)

Step 6: this step is the same as step 3. Add G and the next subkey. The first round
is S[3], and the ith round is [2i + 1]. Starting from 1, the operation result is H
(see Fig. 3.18).

Fig. 3.16 Step 4 of each
round

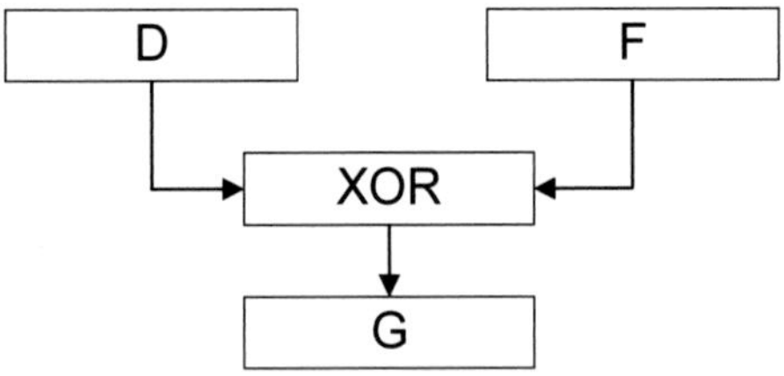

Fig. 3.17 Step 5 of each
round

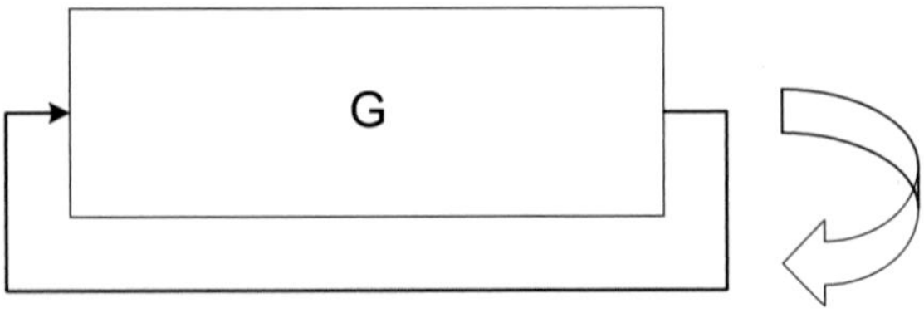

Fig. 3.18 Step 6 of each round

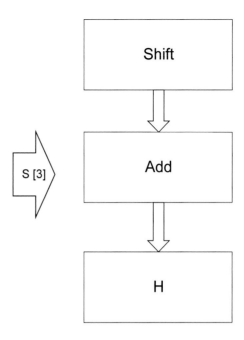

Step 7: this step is mainly to check whether all rounds are completed, execute i + 1 in turn, and check whether i is less than r (if it is less than r, change F to C, H to D, and return to the first step to continue) (see Fig. 3.19).

The generation of subkey in RC5 algorithm is divided into two steps: the generation of subkey and the mixing of subkeys. In the step of subkey generation, two constant P and Q are used, and the generated subkey array is S. The first subkey S[0] is initialized with P, and then each subsequent subkey (S[1], S[2], …) can be calculated according to the previous subkey and constant Q (using 2^{32} modulus). In the step of subkey mixing, the subkeys S[0], S[1], … are mixed with the original key L[0], L[1], …, L[c]. c in L[c] is the part of the last subkey of the original key, which can be seen in Figs. 3.20 and 3.21. Figure 3.20 shows the mathematical form of sub key mixing. Figure 3.21 shows the mathematical form of sub key generation.

Fig. 3.19 Step 7 of each round

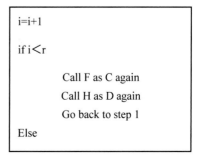

i=j=0
A=B=0

Do 3n times(where n is the maximum of 2(r+1) and c)

$$A=S[i]=(S[i]+A+B)\lll3$$

$$B=L[i]=(L[i]+A+B)\lll A+B$$

Fig. 3.20 Mathematical form of subkey mixing

S[0]=P

For i=1 to 2(r+1)-1
    $S[i]=(S[i-1]+Q) \mod 2^{32}$

Next i

Fig. 3.21 Mathematical form of subkey generation

Because the 56 bit key of DES algorithm is not very secure under the attack of large-scale key search, and the 64 bit block is not strong enough, it is necessary to develop a new algorithm to make up for its shortcomings. Advanced Encryption Standard (AES) is a block encryption standard adopted by the federal government of the United States. AES uses 128 bit block and 128 bit key. According to the designer, the main features of AES are as follows:

(1)   The symmetric and parallel structure makes the implementation of the algorithm very flexible and can resist the attack of cryptanalysis.
(2)   It is suitable for modern processors (Pentium, RISC and parallel processors).
(3)   It is suitable for smart card.
2.   Asymmetric key encryption algorithm.

The biggest advantage of symmetric key encryption algorithm is its high speed in encryption and decryption, which can encrypt a large amount of data, but it is difficult to manage the key and has the problem of key exchange. Because the sender and receiver of encrypted information use the same key in the symmetric key encryption process, the key is easy to be known by others. In view of this shortcoming, asymmetric key encryption algorithm can solve this problem well.

Asymmetric key encryption algorithm needs to use a pair of keys to complete the encryption and decryption operations. One is published publicly, which is the public key. The other is kept secretly by users, which is the private key. The sender encrypts with the public key and the receiver decrypts with the private key. The disadvantage of this algorithm is that the speed of encryption and decryption is much slower than symmetric key encryption.

The differences between asymmetric encryption algorithm and symmetric encryption algorithm are as follows. Firstly, the key values used for message encryption and decryption are different. Secondly, asymmetric key encryption algorithm is thousands of times slower than symmetric key encryption algorithm, but it can protect communication security. Asymmetric key encryption algorithm has the advantage that symmetric key encryption algorithm can't match.

In 1977, Ron Rivest, Adi Shamir and Len Adieman of MIT developed the first important asymmetric key encryption system, called RSA public key encryption algorithm (RSA algorithm), which solved the key agreement and publishing problem. Even today, RSA algorithm is the most widely accepted public key scheme. We only need to publish our own public key to communicate safely on the Internet. All these public keys can be put in a database and can be queried by anyone, but only the private key is known by ourselves.

Asymmetric key cryptography uses two keys to form a pair, one for encryption and the other for decryption. Other keys cannot decrypt this information, including the first key for encryption. The advantage of this design is that each communication party has only one key. Once the key pair is obtained, it can communicate with other people. The mathematical basis of this model is that if a number has only two prime factors, a pair of keys can be generated. For example, the number 14 has only two prime factors 2 and 7. If 7 is used as the encryption factor, then the decryption factor can only be 2, and no other number including 7 can be used as the decryption factor. From this example, it can be inferred that if the number is very large in the design, the cracking process will be very difficult.

One of the two keys is the public key, and the other is the private key. In this mechanism, each party or node will issue its own public key, so that a directory can be formed to maintain each node (ID) and the corresponding public key. By querying the public key directory, any public key can be obtained, so as to communicate with it. Table 3.1 shows the comparison between private key and public key.

The working principle of asymmetric key encryption is as follows.

(1) Because the sender knows the public key of the receiver, when the sender wants to send information to the receiver, it needs to encrypt with the public key of the receiver.

(2) The sender sends the message to the receiver.

(3) The receiver decrypts the message from the sender with its private key. Because the intruder does not know the private key of the receiver, the message can only

Table 3.1 Comparison of private key and public key

Key details	The sender knows	The receiver knows
Private key of the sender	Yes	No
Public key of the sender	Yes	Yes
Private key of the receiver	No	Yes
Public key of the receiver	Yes	Yes

be decrypted with the private key of the receiver. Therefore, no one can get the message except the receiver. Similarly, when the information receiver sends information to the sender, it is the reverse of the above process.

RSA algorithm is the most reliable asymmetric encryption algorithm. RSA algorithm is widely used in public key encryption and e-commerce. In RSA algorithm, the more difficult it is to factorize a maximum integer, the more difficult it is to crack. That is, the more secure and reliable it is. If we find a fast algorithm for large integer factorization, the reliability of information encrypted by RSA cannot be guaranteed. But in fact, it is very difficult to find such an algorithm. Now, only a short RSA key can be broken in a strong way. So far, there is no reliable way to attack RSA algorithm in the world. As long as the length of the key is long enough, the information encrypted by RSA cannot be broken because of the large amount of computation in the current algorithm. RSA algorithm is the first one that can be used for both encryption and digital signature. It is also easy to understand and operate. RSA is the most deeply researched public key algorithm. In more than 40 years since it was proposed, it has experienced various attacks, which reflects the reliability and is gradually accepted by people. It is considered as one of the best public key schemes.

Here is a brief introduction to RSA algorithm. Primes are numbers that can only be divided by 1 and itself. The mathematical basis of RSA algorithm is that it is easy to multiply two large primes, but it is difficult to find the common factor of their product. The private key and public key in RSA are based on large primes with more than 100 bits. Compared with symmetric key encryption algorithm, the principle of RSA algorithm is simple. Its difficulty is to select and generate the private key and public key.

The process of generating private key and public key, encryption and decryption can be summarized into seven steps, which are described as follows:

(1)   Select two large prime numbers P and Q;
(2)   Calculation $N = P \times Q$;
(3)   Select a public key (encryption key) E, so that it is not a factor of $(P - 1)$ and $(Q - 1)$;
(4)   Select the private key (decryption key) D to satisfy $(D \times E) \mod (P - 1) \times (Q - 1) = 1$;

(5)　When encrypting, the equation of calculating ciphertext CT from plaintext PT is $CT = PT^E \bmod N$;

(6)　Send the ciphertext CT to the receiver;

(7)　When decrypting, the equation of calculating plaintext PT from ciphertext CT is $PT = CT^D \bmod N$;

The key of RSA algorithm is to choose the right key. If B wants to accept secret information from A, it needs to generate private key D and public key E, and then send the public key and number N to A. A uses E and N to encrypt the information, and then sends the encrypted message to B. B uses private key D to decrypt the information. As long as the attacker knows the public key E and the number N, it seems that he can find the private key D by trial. At this time, the attacker needs to find the P and Q using N first. But in practical application, when the P and Q are very large, it is not easy to find the P and Q using N. This process is very complex and takes a lot of time. Because the attacker cannot find the P and Q, even if the attacker knows the N and E, he cannot find D. The attacker cannot decrypt the ciphertext. If the symmetric encryption algorithms such as DES and asymmetric key encryption algorithms such as RSA are implemented by hardware, DES will be about 1000 times faster than RSA. If these algorithms are implemented by software, DES will be about 100 times faster than RSA.

Asymmetric key encryption algorithm solves the problem of key agreement and key exchange. Symmetric key encryption algorithm and asymmetric key encryption algorithm have their own advantages and need to be improved. Table 3.2 shows the comparison of symmetric and asymmetric key encryption.

In order to make a good combination of these two encryption mechanisms, aiming at shortcomings to better integrate their advantages, we need to achieve the following goals.

Table 3.2 Comparison of symmetric and asymmetric key encryption

Features	Symmetric key encryption	Asymmetric key encryption
Key used for encryption/decryption	Encryption/decryption uses the same key	Encryption/decryption uses different keys
Encryption/decryption speed	Fast	Slow
The length of ciphertext obtained	Usually equal to or less than the plaintext length	Greater than plaintext length
Key agreement and key exchange	There are big problems	No problem
The number of key needed and the number of information exchange participants	About the square of the number of participants, with fair scalability	Equal to the number of participants with good scalability
Usage	Mainly used for encryption/decryption (confidentiality), not for digital signature (integrity and non repudiation check)	Can be used for encryption/decryption (confidentiality) and digital signature (integrity and non repudiation check)

(1) Fast encryption / decryption speed.
(2) Short length of generated ciphertext.
(3) After combination, the scalability becomes better, but more complexity cannot be introduced.
(4) The solution should be safe.
(5) The key issue needs to be resolved.

The security scheme combining symmetric key encryption and asymmetric key encryption is described as follows (in the scheme, we assume that A is the sender and B is the receiver).

Step 1: The computer of A uses standard symmetric key encryption algorithms such as DES, IDEA and RC5 to encrypt plaintext (PT) and generate ciphertext (CT). The key (K1) used in this operation is called one-time symmetric key, which is discarded after use, as shown in Fig. 3.22.

Step 2: send the one-time symmetric key (K1) to the server, so that the server can decrypt the ciphertext (CT) and recover the plaintext information (PT). Here, a new concept is introduced. A takes the one-time symmetric key (K1) in the first step and encrypts K1 with the public key (K2) of B. This process is called key wrapping of symmetric key. The symmetric key K1 is placed in the logical box and sealed with the public key (K2) of B, as shown in Fig. 3.23.

Step 3: A puts the ciphertext CT and the encrypted symmetric key together in the digital envelope, as shown in Fig. 3.24.

Step 4: At this time, A will send the digital envelope (including ciphertext (T) and symmetric key (K1) wrapped with public key of B) to B through the network, assuming that the digital information contains the above two items, as shown in Fig. 3.25.

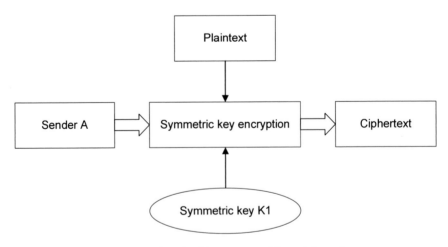

Fig. 3.22 Symmetric key encryption algorithm encrypts plaintext information

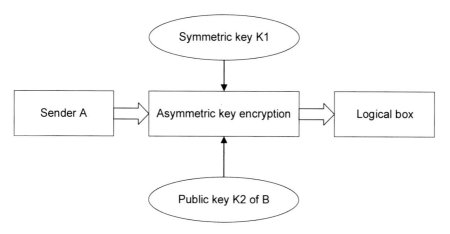

Fig. 3.23 Asymmetric key encryption algorithm encrypts plaintext information

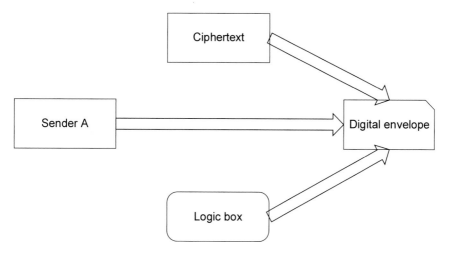

Fig. 3.24 Digital envelope

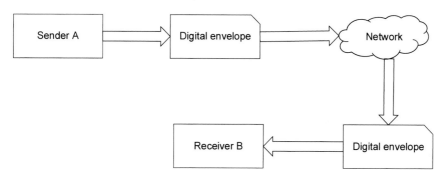

Fig. 3.25 Digital envelope arrives at B through network

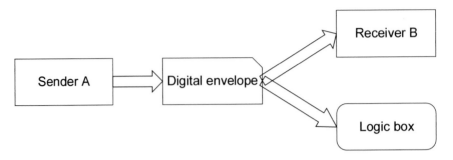

Fig. 3.26 B opening digital envelope with private key

Step 5: B receives and opens the digital envelope. After B opens the envelope, it receives the ciphertext CT and the symmetric key (K1) wrapped with public key of B, as shown in Fig. 3.26.

Step 6: B uses the asymmetric key algorithm used by A and its own private key (K3) to open the logical box. The output of this process is the one-time symmetric key K1, as shown in Fig. 3.27.

Step 7: B decrypts the ciphertext (CT) with the symmetric key algorithm and the symmetric key K1 used by A. This process obtains the plaintext PT, as shown in Fig. 3.28.

The reasons why the digital envelope based process works are as follows:

(1)   We use symmetric key encryption algorithm and one-time session key (K1) to encrypt plaintext (PT). Symmetric key encryption algorithm is fast, and the ciphertext (CT) is usually smaller than the original plaintext (PT). If asymmetric key encryption algorithm is used in this case, not only the speed is slow, but also the output ciphertext (CT) will be larger than the original plaintext (PT);

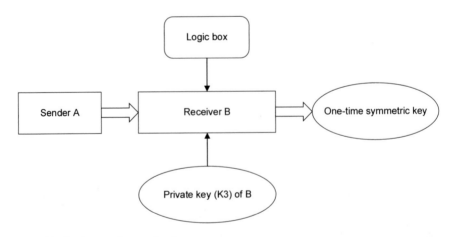

Fig. 3.27 Getting one time session key

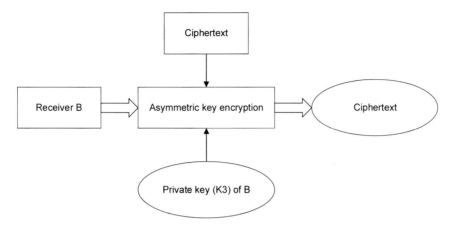

Fig. 3.28 Obtaining plaintext with symmetric key

(2)   Because the length of K1 is small, the asymmetric key encryption will not take too long, and the obtained encryption key will not take too much space. The symmetric key K1 can be encrypted wrapped with public key of B;

(3)   The process based on digital envelope combines the advantages of symmetric key encryption algorithm and asymmetric key encryption algorithm to solve the key exchange problem.

Digital signature, also known as public key digital signature and electronic signature, is a kind of common physical signature written on paper. It uses the technology in the field of public key encryption to identify digital information. The sender encrypts the information with the private key to obtain the digital signature. The process is shown in Fig. 3.29.

Digital signature technology is a technology that encrypts the abstract information with the private key of sender, and then transmits it to the receiver together with the original text. The receiver can decrypt the encrypted digest only with the public key of sender, and then generates a digest for the received text with Hash function, which is compared with the decrypted digest. After comparison, if the information of the two is the same, it means that the received information is complete and has not been modified. Therefore, it is proved that the digital signature can verify the integrity of the information. Digital signature is a process of encryption, and digital signature verification is a process of decryption. The main function of digital signature is to ensure the integrity of information transmission, identify authentication of sender and prevent the occurrence of repudiation in transaction.

It can be seen from the above introduction to the principle of digital signature that it does not solve the problems of slow speed and large ciphertext in asymmetric algorithm. In practical application, the amount of plaintext information may be very large. The usage of the private key of sender to encrypt the whole plaintext information will cause the encryption process to be very slow. Of course, we can consider using

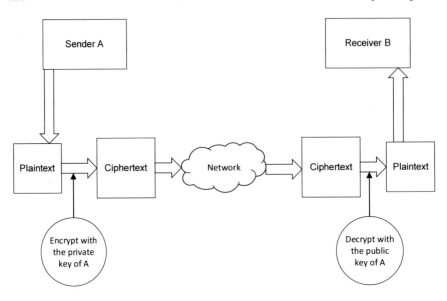

Fig. 3.29 Digital signature

digital envelope to solve this problem. We can also use a more efficient mechanism, which is, using Message Digest, also known as Hash.

Message digest refers to the fingerprint or summary of information, similar to longitudinal redundancy check (LRC) and cyclic redundancy check (CRC). The integrity of data is verified on the premise that the information is not edited after sending and before receiving. Information abstracts usually occupy more than 128 bits. The chance of any two information abstracts being the same is between 0 and 2^{128}. The purpose of this is to narrow the scope of the two information abstracts being the same.

The requirements of message digest can be summarized as follows.

(1) Given a message, it should be easy to get the message digest, and the message digest should be the same, as shown in Fig. 3.30.
(2) Given the message digest, the original message should not be obtained, as shown in Fig. 3.31.
(3) Given two pieces of message, the message digest should be different, as shown in Fig. 3.32.

If two messages get the same message digest, it will violate the above principle, which is called collision. Message digest algorithm usually produces 128 bits or 160 bits of digest information. The probability of two message digests being the same is $1/2^{128}$ or $1/2^{160}$ respectively, which is very unlikely in practice. Even if there is only a slight difference between the two information, there will be differences in the message digest, from which we cannot see the similarity of the two messages. For a message and its message digest, the message digest should ensure that it will not

Fig. 3.30 Giving an
message

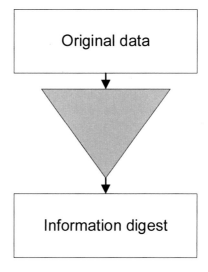

Fig. 3.31 The message
digest cannot be obtained in
reverse

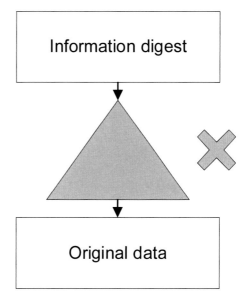

find the message digest with the same message to the maximum extent. The message digest cannot expose the original message, as shown in Fig. 3.33.

MD5 message digest algorithm was designed by American Cryptologist Ronald Linn Rivest and published in 1992 to replace MD4 algorithm. It is a widely used cryptographic Hash function. In order to ensure the integrity and consistency of information transmission, it can generate a hash value of 128 bits (16 bytes). It adds the concept of "safety belts" to MD4, thus, it is safer than MD4. In MD5 algorithm,

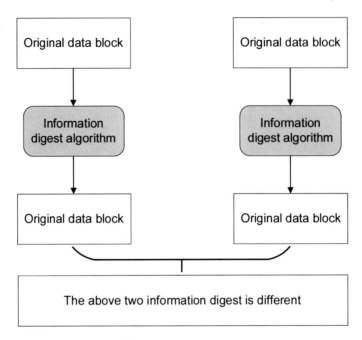

Fig. 3.32 Different message has different message digest

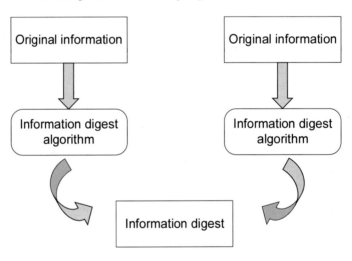

Fig. 3.33 The message digest cannot expose the original message

the size of message digest and the necessary conditions for filling are exactly the same as MD4.

We all know that everyone on earth has their own unique fingerprint, which is a safe and effective way to distinguish different individuals. Similarly, we can apply MD5 to generate an equally unique "digital fingerprint" for any file, regardless of the size, format and number of files. No matter what changes anyone makes to the file, its MD5 value ass the corresponding "digital fingerprint", will change. The purpose of the MD5 value seen in some software download sites is that after downloading the software, we can check the MD5 value of the download file once to ensure the consistency of the download file and the files provided by the site.

From the above description, we can see that the MD5 value is like the "digital fingerprint" of a specific file. The MD5 value of each file is different. If someone changes the file, its MD5 value will also change. For example, the download server provides an MD5 value for a file in advance. After downloading the file, the user can recalculate the MD5 value of the downloaded file to determine whether the downloaded file is correct or not. MD5 algorithm is widely used in software download station, forum database, system file security and other fields.

The working principle and process of MD5 are described as follows.

Step 1: adding fill bits to the original information to make the length of the original information equal to a value as 64 bits less than the multiple of 512, so that the length of the original information after filling is 448 bits, 960 bits, 1472 bits, etc., as shown in Fig. 3.34.

Step 2: after adding the fill bit, first we calculate the length of the information, excluding the length before adding the fill bit. Then we calculate the length of the original information excluding the fill part, and add it to the back of the information and the fill bit, which is represented by 64 bits. If the length of the information exceeds 64 bits, only the last 64 bits will be used. After adding, it is the final information, that is, the information to be hashed. In this way, the length of the information is a multiple of 512.

Step 3: we divide the input into 512 bit blocks, as shown in Fig. 3.35.

Step 4: we need to initialize four link variables, which are called A, B, C and D. They are all 32-bit numbers. The initial hexadecimal values of these link variables are shown in Table 3.3.

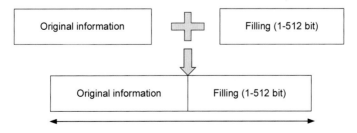

Fig. 3.34 Filling process

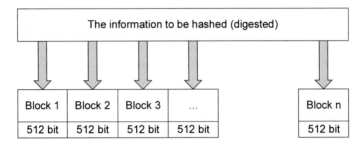

Fig. 3.35 Divide the input into 512 bit blocks

Table 3.3 Link variables

A	Hexadecimal	01	23	45	67
B	Hexadecimal	89	AB	CD	EF
C	Hexadecimal	FE	DC	BA	98
D	Hexadecimal	76	54	32	10

Step 5: After initialization, the algorithm starts the cyclic operation of multiple 512 bit blocks in the information.

First, copy the four link variables into the four variables a, b, c, d, so that a = A, b = B, c = C, d = D, as shown in Fig. 3.36. In fact, this algorithm combines a, b, c, d into 128 bit registers. The registers (a, b, c, d) store the intermediate result and the final result in the actual algorithm operation, as shown in Fig. 3.37.

Then, the current 512 bit block is decomposed into 16 sub blocks, and each sub block is 32 bits, as shown in Fig. 3.38.

Finally, four rounds of processing are needed. Each round processes 16 sub blocks in a block. In each round, the input is as follows: ① 16 sub blocks ② variables a, b, c, d ③ constant t. Each round has 16 input sub blocks M[0], M[1], …, M[15], or M[i], where I is 1 to 25, each sub block is 32 bits. t is a constant array, which contains 64 elements, each element is 32 bits, and the elements of the array t are expressed

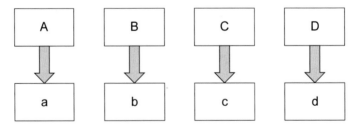

Fig. 3.36 Copy 4 linked variables to 4 variables

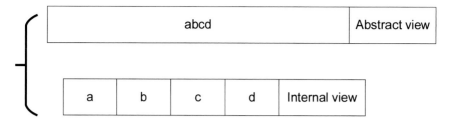

Fig. 3.37 Link variable abstract view

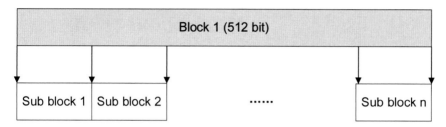

Fig. 3.38 Decomposes the current 512 bit block into 16 sub blocks

as t[1], t[2], …, t[64], or t[k], where the value is 1 to 64, and each round is evenly distributed with 16 t values (see Fig. 3.39).

We summarize these four rounds of iterations. Each round has 16 registers, and its output intermediate and final results are copied to registers a, b, c and d (each round has 16 registers).

Step 1: process P is first to deal with b, c, d which are different in four rounds.
Step 2: add variable a to the output of process P.
Step 3: the information sub block M[i] is added to the output of step 2.
Step 4: add constant t[k] to the output of step 3.
Step 5: the output cycle of step 4 is shifted to the left for S bits.
Step 6: add variable b into the output of step 5.

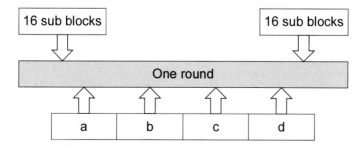

Fig. 3.39 Each round of processing

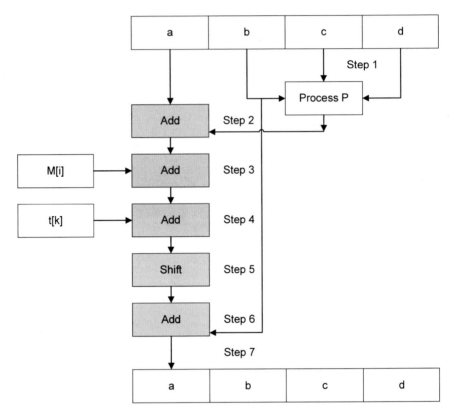

Fig. 3.40 MD5 operation process

Step 7: the output of step 6 becomes the new a, b, c, d of next step.

Figure 3.40 shows the operation of MD5.

MD5 is also widely used in operating system login authentication, such as UNIX, all kinds of BSD system login password, digital signature and many other aspects. For example, in UNIX system, the user password is stored in the file system after hash operation with MD5 or other similar algorithms. When the user logs in, the system uses MD5 hash operation to calculate the password entered by the user, and then compares it with the MD5 value in the original file system to judge whether the password is correct or not. The advantage of this method is that it can avoid the user password being known by the system administrator, and enhance its reliability. Even if you see the description of the source program and algorithm, you cannot convert an MD5 value back to the original string. This is because MD5 can map a "byte string" of any length to a 128 byte large integer. Because there are infinite original strings, we cannot deduce the original string through the 128 bytes. There is no reversible operation.

Other algorithms such as elliptic curve encryption and Elgamal technology are becoming more and more popular in recent years. Usually in data transmission, both MD5 and SHA1 algorithms need to send and receive data. Both sides know the key generation algorithm before data transmission. Unlike HMAC, a key needs to be generated. The sender uses the key to digest the data (generate ciphertext), and the receiver uses the key to digest the received data, and then judges whether the generated ciphertext is the same.

3.2 Encryption Chip

3.2.1 Overview of Encryption Chip

For IOT applications, in addition to the aforementioned soft encryption using encryption algorithm, encryption chip can also be used as a hard encryption means. Encryption chip is a kind of chip with high security level. It can integrate various symmetric encryption algorithms or asymmetric encryption algorithms, so as to ensure that the stored key and information data will not be illegally read and tampered.

In the field of copyright protection, encryption chips can be divided into two categories according to different encryption schemes and usage. Among them, authentication encryption chip is widely used. Its advantage is that the encryption algorithm is unified, secure and simple in practical application. However, the overall encryption scheme of authentication encryption chip has low security and weak protection, which makes it have security risks and cannot better deal with the attack against the main control MCU on the board. Another type of encryption chip uses algorithm and data migration scheme, aiming at the missing function of MCU, and on the premise of ensuring the safe operation of the program, transplants part of the program and data of the main control MCU on the board to the encryption chip, so as to ensure and improve the overall security.

The storage space of encryption chip is 32–128 KB, and the process precision can reach 60 nm or higher. In view of the weakness of traditional logic encryption chip, such as weak protection ability and easy cracking, a smart card platform with good security has emerged. In order to ensure the high security of the smart card chip, it is necessary that the smart card chip has EAL4 + level or above of the International Security Certification Council, otherwise the security is difficult to meet the requirements.

3.2.2 The Function and Basic Principle of Encryption Chip

Only the encryption chip system certified by PBOC has high security and reliability. In order to achieve the purpose of protecting the results, the encryption chip can protect the program burned into flash, so that it cannot run on the illegal board even when it is stolen. On the basis of considering the security performance of the chip operating system, we should also effectively manage the internal resources of the chip, improve the protection ability of the underlying interface, and then improve the security performance of the chip operating system to prevent the system from being attacked or cracked.

1.   The working principle of encryption chip

In the following process, we take AT88SCxx series chip and AT88SC0104C chip as examples. The AT88SCxx series encryption memory chip has many applications. The chip uses serial bus communication in application, and uses authentication or encryption verification to access data. This series of chips are widely used because of its small size, large storage capacity, safety and reliability. AT88SCxx series chips include AT88SC0104C, AT88SC0204C, AT88SC0404C, AT88SC0808C, AT88SC1616C, AT88SC3216C, AT88SC6416C, AT88SC12816C, AT88SC25616C and other models. The user storage space is 1 KB to 256 KB.The package of AT88SC0104C is shown in Fig. 3.41.

A 64bit encryption algorithm is built into AT88SC series chip, which is used in its authentication and encryption mode. The built-in encryption algorithm not only enhances the access security of this series of chips, but also can restrain the bypass attack. In these two modes, the interaction information between host and chip is different, which increases the difficulty of stealing effective information.

AT88SCxx series chips work in three modes: standard, authentication and encryption. In the standard mode, the chip can read and write data according to its own commands. The chip can be accessed as a commonly used erasable programmable read and write memory. Compared with the standard mode, the authentication mode is much more complex. Here first introduces the algorithm of encryption chip, which is called F2 algorithm.

Fig. 3.41 Package of AT88SC0104C

The input of F2 algorithm is a number Q_0 randomly generated by the system, an 8-byte ciphertext C_I read out from the encryption chip and a seed G_C. There are four groups of user access areas in the chip. The output of the algorithm is a new 8-bit ciphertext Q_1. When the host starts to call the F2 algorithm, it sends and with the verify authentication command to the AT88SCxx chip. After receiving the authentication command, it performs the same operation to generate Q_2 and key S_K according to the 8-byte ciphertext C_I and seed G_C read out from the encryption chip, and compares Q_2 with Q_1. If the authentication is successful, after replace Q_2 with Q_1, a new ciphertext is generated to update the configuration area. In encryption mode, the data transmitted by bus is encrypted ciphertext. This mode requires the user to pass the authentication once. After the authentication is successful, the user has the updated data in the specific register of the configuration area. Then the data is used as the key for the second authentication. Finally, when accessing the user area, the user needs to pass the command verification set by different user areas, and the authentication can be carried out after matching.

Further subdivide the AT88SCxx series, for example, the AT88SC0104 chip is divided into four data storage areas, each of which has a capacity of 32 bytes, and the size is $4 \times 32 = 128$ B. After the partition, the register is configured. After setting the AR register, the user area can be read and written through the authentication command, read and write data, and send the command of checksum.

2.   The use stage of encryption chip

The use stage of encryption chip includes three stages.

1)   Development stage

The development stage is the first stage of the encryption chip using. In this stage, we need to debug the code very carefully. Because in this stage, when the number of accesses to some registers exceeds their count times, it is likely to cause the chip to be locked, resulting in the failure of the unlock configuration area.

2)   Fusing stage

When the configuration of the development stage is finished, the second stage of using the encryption chip is started. The chip is fused with the fusing command. At this stage, we need to pay attention to confirm whether the custom algorithm can calculate the unique value G_C.

3)   Factory stage

The factory stage is the last stage of using encryption chip. The main work of this stage is to solder the encryption chip to the board.

3.   Interface description of encryption chip.

For the application layer, we can only see the following four interfaces.

```
int SE_Load_Data(int zone, puchar data, uchar len);
int SE_Save_Data(int zone, puchar data, uchar len);
```

int SE_Auth_Done();
int SE_Auth_Init();

where, the first two functions use encryption authentication, read and write user area data, encrypt timing. The third function is an authentication function. In each call process, the host will initiate an encryption authentication operation. The main work of the fourth function is to complete the initialization of hardware GPIO, testing the timing and type of the chip.

3.2.3 Security Protection of Encryption Chip

We take DX81C04 encryption chip as an example, DX81C04 encryption chip is shown in Fig. 3.42.

1. The security of encryption chip itself

There is a random number generator inside the encryption chip, which will have different answers to the same question each time. This mechanism ensures the security performance of the encryption chip when dealing with different situations, and ensures that the data can be measured online but cannot be simulated.

The calculation factor in the encryption chip can ensure that the key written to each chip is different and cannot be copied.

2. Product design

When using the encryption chip DX81C04, we should pay attention to hiding the truth value comparison point in the program to improve the difficulty of cracking the program.

3. Users

Fig. 3.42 Dx81C04 encryption chip

The key of DX81C04 encryption chip includes real key and sub key. The real key is in the hands of the key person, and the programmer only uses the sub key in the application. It should be noted that the sub key cannot burn the chip. Therefore, it is difficult for other people to obtain the real key.

4. Production

Encryption chip DX81C04 has its own unique key control method. In the process of key burning, this method uses encrypted USB key to ensure the security of key and prevent key leakage, so that the burner can burn the chip safely. This method is very safe and controllable in practical application. In the production process, key burning is an important part of encryption chip.

5. Security protection of chip users

Before leaving the factory, the chip is defined with a unique CID number, so that the library function and USB key of each user are unique for special use. That is to say, each encryption chip has uniqueness to prevent the collusion and plagiarism between encryption chips.

3.2.4 Defense System of Encryption Chip

1. Application fields of encryption chip

All kinds of IOT products using encryption chips: Bank encryption U shield, recorder, encryption hard disk, PC Lock, mobile phone, intelligent door lock, sensors of key nodes of underground pipe network, etc. When using these products, because the data stored in these encryption chips have been processed reliably, they are difficult to be stolen or embezzled, which ensures the security of data information.

The user data that needs to be processed can be encrypted with the key, and then stored in the encryption chip. For example, the initial power-on password, hard disk password, fingerprint information, and key instructions used for data encryption on the hard disk will be encrypted again and stored in the security chip. In this way, if you want to crack it, you must first crack the encryption chip.

Usually, the hard disk encryption can be completed as long as the password is set. This setting is simple and convenient for users, but once the password is forgotten, it means that the hard disk cannot be used. Similarly, if other people get the hard disk without knowing the password, they cannot read the data in the hard disk, which protects the data stored in the hard disk to a certain extent.

2. Processing speed of encryption chip

Another important index of encryption chip is the processing speed of encryption and decryption. Taking 3DES algorithm as an example, the data stream encryption implemented by MCU software is very slow, but the processing speed of 6 Mb/s can be easily achieved by using encryption chip. Another example is the more complex

Table 3.4 The running speed of FM15160 encryption chip

Run the algorithm	FM15160@CPU = 24 MHz, RAE = 48 MHz
SSF33	16 Mb/s
SM1	8 Mb/s
DES/TDES	6 Mb/s
AES	1 Mb/s (Soft implementation)
SMS4	950 kb/s (Soft implementation)
SM7	9 Mb/s
RSA generates key pair	1024: 1.2 s/time (without CRT/with CRT) 2048: 16.5 s/time (without CRT/with CRT)
RSA signature	1024: 80 ms/time (without CRT) 1024: 26 ms/time (with CRT) 2048: 550 ms/time (without CRT) 2048: 150 ms/time (with CRT)
SM2	12 ms/time
SM2 signature verification	38 ms/time
SM2 encryption	39 ms/time
SM2 decryption	28 ms/time
ECDSA signature	192 bit: 18 ms/time 384 bit: 77 ms/time
SM3	25 Mb/s
SHA1	25 Mb/s
SHA256	25 Mb/s
MD5	3 Mb/s (Soft implementation)

ECC algorithm. On the A1006 encryption chip, the authentication time is 50 ms. Table 3.4 shows the operation speed of FM15160 encryption chip.

3. Security features of encryption chip

    (1) The chip tamper proof design makes the encryption chip have a unique serial number, which can prevent SEMA/DEMA, SPA/DPA, DFA and timing attacks.

    (2) A variety of detection sensors: high and low pressure sensors, frequency sensors, filters, pulse sensors, temperature sensors, etc.

    (3) The built-in sensor of the encryption chip has the function of life test, which can carry out the self destruction function of the chip against the external illegal detection.

    (4) Bus encryption makes the encryption chip have a metal shielding layer. After the encryption chip detects the external attack, it performs the internal data self destruction function.

    (5) The encryption chip has a true random number generator, which can avoid the generation of pseudo-random number.

4. Evaluation guarantee level of encryption chip

The evaluation guarantee level of encryption chip must follow the grading evaluation criteria of information security products. Classified evaluation of information security products is based on the national standard GB/T18336-2001. It comprehensively estimates the application environment of the target product, and also comprehensively evaluate and test the whole life cycle of information security, including technology, development, management, delivery and other parts, verify the confidentiality, integrity and availability of the product. After that, it determines whether the product is safe enough for the expected application and can withstand the predetermined risks in the use process. Finally, the product can meet the corresponding assessment assurance level.

The following seven assessment assurance levels are defined in GB/T 18336-2001.

(1)   Evaluation assurance level 1 (EAL1)— Function test.
(2)   Evaluation assurance level 2 (EAL 2)—Structural testing.
(3)   Evaluation assurance level 3 (EAL3)—System testing and inspection.
(4)   Evaluation assurance level 4 (EAL4)—System design, testing and review.
(5)   Evaluation assurance level 5 (EAL 5)—Semi formal design and testing.
(6)   Evaluation assurance level 6 (EAL 6)—Design and test of semi formal verification.
(7)   Evaluation assurance level 7 (EAL7)—Design and testing of formal verification.

Hierarchical evaluation obtains the security assurance level through the independent evaluation of the security of information technology products, indicating the security and credibility of the products. The higher the level of authentication, the higher the security and credibility. The product can resist higher level threats, suitable for high risk environment.

3.3   Cloud Platform

3.3.1   The Concept of Cloud Platform

1.   Cloud platform overview

Cloud platform, also known as cloud computing platform, can be used as the background for the analysis, processing and decision-making of IOT data. Cloud platform can be divided into three categories, namely storage cloud platform, computing cloud platform and integrated cloud platform. Among them, the storage cloud platform is mainly focused on data storage, the computing cloud platform is mainly on data processing, and the integrated cloud platform is on both computing and storage. The emergence of a variety of cloud platforms promotes the cloud platform to cloud computing. The so-called cloud platform is a platform that allows developers to run programs in the "cloud" or use services in the "cloud" for other visitors. As a new way to support applications, cloud platform has great potential. On demand platform and platform as a service (PaaS) are also the names of cloud platform.

2.   Cloud services

In the actual environment, there are three kinds of cloud services, namely software as a service (SaaS), attached services and cloud platforms.

1)   Software as a service

SaaS application runs completely in the "cloud", which can correspond to the server of an Internet service provider. Its on-premises client is usually a browser or other simple client. At present, the most famous SaaS application is Salesforce. The service provider can provide an on-demand customer relationship management platform, allowing customers and independent software suppliers to customize and integrate their products, and establish their own application software.

2)   Attachment service

Attached service means that every on-premises application can access the services provided by "cloud" through some specific applications. These functions of indoor applications need to be attached to these applications. For example, iTunes of Apple Inc., whose desktop application can be used to play music, etc., will enable users to purchase new audio or video. For another example, Microsoft Exchange hosting service is an enterprise level example. It can add "cloud" based spam filtering, archiving and other services to indoor exchange servers.

3)   Cloud platform

The direct users of cloud platform are developers. When developers create applications, they can use "cloud" services based on cloud platform. On the cloud platform, developers only need to rely on the cloud platform to create new software as a service applications, which saves the time and energy of building the basic framework.

3.   Cloud platform infrastructure

Cloud platform can achieve on-demand access to users. By virtualizing physical resources into virtual machine resource pool, it can help visitors flexibly call the required software and hardware resources, so as to achieve on-demand access to users. In the actual operation process, the cloud platform migrates the virtual machine resources in real time according to different user concurrency, so as to minimize the resource cost, improve the utilization of CPU and memory, and ensure high-quality service.

The architecture of cloud platform is mainly divided into four layers, namely resource layer, virtual layer, middleware layer and application layer. The structure and function of each layer are described below.

1)   Resource layer

The resource layer consists of server clusters. Traditional servers need high performance servers with large memory, fast CPU speed and large disk space to provide high-quality services. The cost is expensive. The server cluster can overcome this shortcoming by using distributed processing technology to combine servers with poor performance to provide reliable services.

2)   Virtual layer

With the resource layer, in order to minimize the resource cost and maximize the resource utilization, it is necessary to establish a virtual machine. If the memory of a physical machine is small and a small batch of tasks need to be processed continuously for a period of time, several virtual machines can be set up on the physical machine to process the application requests. At this time, each virtual machine is equivalent to a small server. Hypervisor, or KVM for short, can be used as the switch of virtual machine and can also allocate resources to virtual machine. In order to form a virtual machine pool, we use KVM to allocate an appropriate amount of memory, CPU, network bandwidth and disk to each virtual machine.

3)   Middleware layer

Middleware layer is the core layer of cloud platform architecture, which can monitor, warn and optimize the resource status of virtual machine pool. The middleware layer has the following functions.

(1)   Monitoring: It monitors the real-time usage status of each virtual machine, including the usage status of CPU and memory in the virtual machine. It also monitors the application requests of users, so as to make better decisions based on the above monitoring results.

(2)   Early warning: before making resource adjustment, it predicts the amount of user requests according to the current usage of virtual machine resources, so as to prevent the occurrence of problems. For example, when it is predicted that the CPU utilization limit will be reached in the next second, the corresponding response should be made.

(3)   Optimization decision: virtual machine needs to migrate or scale resources in this process. We focus on the utilization of resources in this process, and ensure the quality of service. At the same time, it also determines what kind of scheduling strategy we should adopt. In this layer, we need to build a load balancing system, operating system, file storage system and server to respond to the application and realize the function of middleware layer. This process is just like that the router only forwards the data without data processing. The data processing is performed by the Tomcat server on the virtual machine.

When the whole cloud platform runs for the first time and receives the user request, if the user scale is large and the virtual machine needs to process tasks one by one, an appropriate allocation policy needs to be found in the virtual machine to minimize the cost and maximize the quality of service. This is another function existing in the middleware layer, that is, the initialization allocation function.

4)   Application layer

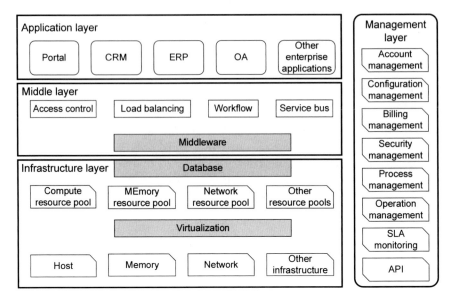

Fig. 3.43 Cloud platform infrastructure model

The application layer is mainly to provide users with a visual interface. The application is divided into two categories: storage and application server. If the application is storage, such as online storage service, it will provide the user with an interactive interface, on which the user can create a folder for data storage, and has the function of playing video online. If the application is a leased server, the visual interface will have the resource status of the leased server.

The cloud platform infrastructure model is shown in Fig. 3.43.

3.3.2   Characteristics of Cloud Computing

1.   Definition of cloud computing

In August 2006, Eric Schmidt, chairman of Google, first proposed the concept of cloud computing at SES San Jose 2006. Since then, the word "cloud computing" has rapidly occupied the major IT news, and also affected the global stock market. The most common definition is to divide the cloud computing industry into three levels, namely cloud software, cloud platform and cloud devices, corresponding to software as a service (SaaS), platform as a service (PaaS) and infrastructure as a service (IaaS).

Cloud computing in cloud technology is usually classified as data center products. Distributed storage technology is used in the cloud to provide services to front-end users. The services used by users are the result of tens of thousands of computers running together in the cloud. Cloud technology can be regarded as a subset of network technology. Both of them have the same purpose, which is to hide the complexity of the system, so that users can use it without knowing how the system works inside. Cloud computing is developed from the distributed parallel computing technology and concept of network technology. It uses new terms to package the traditional technology, and its essence is unified. Cloud computing was initially positioned as a commercial use, hoping to change the application mode, which is the biggest difference between cloud computing and all the technologies before it. Cloud computing not only requires service providers to provide services and charges according to the needs of different users, but also enables users to experience the convenience, quickness and benefits of cloud computing.

2. Features of cloud computing
1) Simplified architecture

Compared with the complicated problems in network computing, such as the compatibility of different servers, different operating systems and different compiler versions, cloud computing has a simpler architecture. Taking cloud computing of Google as an example, it does not need to solve different problems. It uses a large number of personal computer level servers with the same specifications to execute programs, and uses the parallel computing system architecture to help coordinate the information transmission between servers, so as to optimize and improve the performance of distributed processing in network computing as a whole.

2) Gradually improvement

In practical applications, the number of servers presents a dynamic growth trend, which requires that the infrastructure of cloud computing also needs to grow with its growth. Although the computing and storage capacity of cloud center will not be in place in one step, it can gradually increase with the demand. In this way, providers of cloud services can avoid high investment risks.

3) System failure is normal

The construction of cloud computing needs to consider the reduction of the number of servers due to system failure. But one of the characteristics of cloud computing is that the system failure is regarded as a very normal situation, which is very different from the usual parallel computing that server failure will lead to serious consequences.

4) Network support

In cloud computing, if we want to show the advantages of parallel computing, so that it is not limited by network bandwidth and other factors, we need a strong network to support the parallel computing between servers and the information transfer between cloud services and users. This is because when we transfer the results to end users in cloud computing, we should first consider the quality of the results, and then speed up the delivery on the premise of ensuring the quality.

5) Virtualization technology

The architecture of parallel computing makes the end-user need not know the number of back-end servers in the whole architecture. After the request is sent, the end-user can get the expected computing power and storage space. The result is virtualization.

6)   Tasks with less data

Cloud computing focuses on small tasks such as executing single data processing.

7)   Distributed file system

The requirements supported by cloud computing include both computing intensive and data intensive. It stores the computing and data in the same node. With distributed file system, the data that each node needs to obtain can be known from the client without increasing the network access space.

8)   Map/Reduce

When dealing with software programs with a large amount of data, the function of Map is to cut the data so that they are not related to each other, and then send them to the computer for corresponding distributed processing. The function of Reduce is to merge the results of distributed processing and output them, and take the output of Reduce as the final result.

9)   Provide perfect development environment

According to the previous introduction, cloud computing is initially positioned as a commercial use. One of the commercial models it builds is to enable enterprise users to run their own developed system platform. The purpose of doing so is not only to ensure the provision of resources on the Internet, but also to strengthen the management of servers and operating systems. For example, the support for API even provides a new programming language.

Figure 3.44 shows the cloud computing platform.

3.4   Information Fusion

3.4.1   Theoretical Concept of Information Fusion

1.   Origin and definition

The term "information fusion" originated from the sonar signal processing system funded by the U.S. Department of defense in 1973. It initially appeared in the name of data fusion. In the 1980s, multi-sensor data fusion (MSDF) technology appeared to meet the needs of military operations. In the early 1990s, with the development of information technology, the concept of more comprehensive "information fusion" was gradually incorporated. In 1988, the data fusion technology in C^3I (command, control, communication and intelligence) system was listed as one of the 20 key technologies developed by the Department of Defense. In the Gulf War, the application

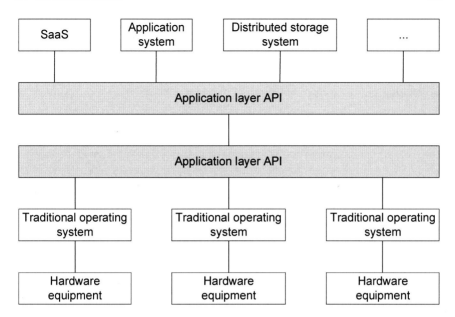

Fig. 3.44 Cloud computing platform

of information fusion technology shows great potential and operational superiority. After the end of the war, the U.S. Department of Defense added computers to C^3I system to further improve its performance. Since then, C^3I system with information fusion as the core was born. Information fusion technology originated in the field of military applications. With continuous innovation and development in this field, its scope of application continues to expand. Related technologies continue to update, in mathematics, military science, computer science, automatic control theory, artificial intelligence, communication technology and other fields have cross applications. Now it has been widely used in civil engineering.

From the military point of view, information fusion can be understood as multi-level, multi-faceted and multi-level processing of multi-source information and data. Data from multiple sensors or other information sources are fused to obtain more accurate state and type judgment, so as to carry out more complete situation and threat assessment. Its processing methods include detection, correlation, correlation, estimation and synthesis. The U.S. Department of Defense defines data fusion from the perspective of military applications. Waltz and others have supplemented and modified this definition, and given a more complete definition of information fusion. Information fusion is a multi-level and multi-faceted processing process. This process is to detect, combine, correlate, estimate and combine multi-source data to achieve accurate state estimation and identity estimation, as well as complete and timely situation assessment and threat assessment. We can also think that the purpose of information fusion is to accurately locate and estimate the target identity, to evaluate the real-time battlefield situation and enemy threat, and to interconnect and evaluate the data from single or multiple information sources. This process can be divided into alignment, interconnection, identification, threat assessment and battlefield situation assessment.

Alignment is to arrange data in time, space and measurement units. Multi sensor information fusion needs to transform all the data collected by different coordinate systems, observation time and scanning period into a common reference system, and then process the data. The most typical method is least square estimation.

Interconnection is correlation, which is simultaneous interpreting the measured data from different sensors and selecting the data for tracking the same target. Data interconnection can be carried out at three levels: measurement-measurement interconnection, which is mainly used to process the initial tracking of a single sensor or system; measurement-tracking interconnection, which is mainly used to track and maintain; tracking-tracking interconnection, which is mainly used to process multi-sensor data.

Identification is to estimate the attributes of the target. First we identificate the shape features such as shape and size, and then deal with the problem of decision theory, which can be divided into Bayesian reasoning method and D-S evidence theory method. At present, Bayesian reasoning method is widely used.

Threat assessment and battlefield situation assessment is a relatively complex process, which needs a large number of databases to complete. The database must include different target behavior, target flight trend, future attempt data and intelligence information of enemy military forces. Bayes method, D-S evidence theory and artificial intelligence technology can be applied to this level.

In a word, information fusion is an information processing technology that takes multi-source information as the processing object. It uses computer technology to automatically analyze and comprehensively process the information detected by multi-sensor from the same or different classes according to certain rules. It continuously modifies and refines the processing results, so as to automatically generate the desired synthetic information.

2.   The arrangement of sensors in information fusion

Sensor placement is the most important problem in the sensor fusion structure, which can be divided into three types: parallel topology, serial topology and hybrid topology (tree topology). In the parallel topology, all kinds of sensors work at the same time. In the serial topology, sensor detection is temporary, such as SAR (Synthetic Aperture Radar) images belong to this structure.

3.   The structure of fusion level

Information fusion can be carried out before and after the information obtained by each sensor is preprocessed or after the sensor processes the components and completes the decision. According to the processing process of the data before being sent to the fusion center, data fusion can be divided into data level fusion, feature level fusion and decision level fusion.

The data layer fusion is the lowest level fusion. In the fusion process, the original information sent by each sensor to the fusion center is fused. The original information

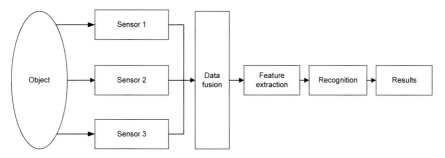

Fig. 3.45 Data fusion process

sent must be matched, which is not processed or rarely processed. Figure 3.45 is the data layer fusion process. The fusion of data layer requires that in the fusion process, all the sensors participating in the fusion should be accurate to the registration accuracy of one pixel when transmitting information. The fusion can be carried out on pixels or resolution units. The pixels can cover one-dimensional time series data, focal plane data, etc. The advantage of data layer fusion is that it can save as much useful information as possible, and provide some subtle information that can only be provided in this layer. The disadvantage of data layer fusion is that it needs to deal with a large amount of data in this layer, which is time-consuming, poor real-time, and the information obtained is more uncertain and unstable, with poor information stability, large data traffic and poor anti-interference ability.

Compared with the other two levels, feature level fusion can be called "intermediate fusion", which can be divided into target state information fusion and target feature fusion. It uses the original information provided by each sensor, first extracts a group of feature information to form a feature vector, and then fuses each group of information to classify the target or do other related processing. The extracted feature information should be a sufficient representation or a sufficient statistic of pixel information. Finally, classification, aggregation and synthesis are carried out according to the extracted feature information. Figure 3.46 shows the fusion process of feature layer. In this layer fusion, it takes the advantages of data layer and decision layer

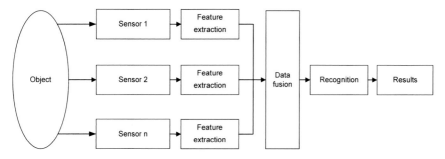

Fig. 3.46 Feature layer fusion process

fusion into account. In the three levels of fusion, feature layer fusion is developed the most. Feature level fusion achieves considerable information compression and facilitates real-time processing of information. The features extracted in the fusion process are directly associated with the final decision analysis. Therefore, the feature information needed for decision analysis can be given to the maximum extent. But in this layer, because the feature vectors are extracted from the observation data and then connected to a single vector, the communication bandwidth is reduced, and the information loss reduces the accuracy of the results. Decision level fusion is carried out at the highest level. In decision level fusion, each sensor makes corresponding decisions based on its own single source data. The fusion of these results becomes the final decision, which is also the optimal decision. Figure 3.47 shows the decision level fusion process. Because its fusion output is the result of a joint decision, theoretically, the joint decision should be more accurate than any single sensor decision. The methods used in the decision-making layer mainly include Bayesian inference, D-S evidence theory, fuzzy set theory and expert system theory. There are many successful applications in the fields of threat recognition alarm system, multi-sensor target detection, industrial process fault detection and robot vision processing on the tactical aircraft platform. The advantage of decision level fusion is that the fusion has good fault tolerance, and it can work even when there is a sensor failure. The information fusion has less traffic, strong anti-interference ability and strong real-time performance.

4.   The advantages of information fusion technology

1)   Providing stable performance

In the system, each sensor can provide target information independently without mutual interference. When one of the sensors is interfered and cannot work normally, the other sensors can continue to work without its influence.

2)   Improving spatial resolution

Multi sensor can use sensor aperture to obtain higher resolution than any single sensor, which can greatly improve the overall spatial resolution.

3)   Obtaining more accurate target information

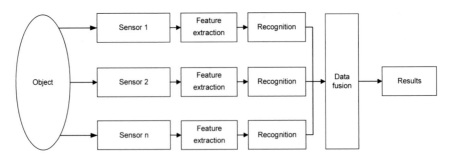

Fig. 3.47   Decision level fusion process

Using the information provided by multi-sensor can improve the accuracy of target event estimation. The independent measurement of different sensors on different timing of the same sensor or simultaneous interpreting of different sensors can improve the detection performance and improve the credibility of the results.

4)   Obtaining the target information that single sensor cannot get

Multi sensor information fusion can avoid the limitations of a single sensor, obtaining more and more effective information, to improve the ability of target recognition. The frequency complementarity between multiple sensors can expand the coverage of time and space, and the dimension of the measured space will also increase with the increase of the number of sensors, reducing the detection blind spots caused by external reasons such as concealment, deception, camouflage and meteorological terrain. In the multi-sensor system, for the same detection object, the comprehensive detection information of the object can be obtained by using various sensor detection information and different processing methods, so as to improve the detection accuracy and reliability.

In command decision-making, information fusion technology can increase the accuracy and reliability of judgment, and reduce the cost of weapon system in many cases. In combat, appropriate allocation of sensors can improve the efficiency of detecting and tracking more enemy targets.

5.   Time fusion and space fusion

Time fusion is to fuse the observations of the target at different times. It should be noted that the fusion should be carried out according to the order of observation time. Space fusion is to fuse the observations of different positions at the same time, which is suitable for multi-sensor fusion. In practical application, in order to get better comprehensive estimation, the two methods are often combined.

When observing the moving target, the observation values of sensors distributed in different positions will be different in different time and space, thus, forming a set of observation values.

Beidou satellite navigation system can provide accurate time and space information, which is the most convenient and accurate means for information fusion and space–time alignment with IOT sensors.

3.4.2 Basic Fusion Methods

1.   Data support of information fusion

In order to manage a given decision task with information fusion method, we need to use multi-source information for data support according to the task requirements, which involves many aspects. Firstly, we need to deploy multiple sensors related to the decision task, collect information resources related to the task, and combine the above information for future research. Secondly, in order to meet the needs of goal and time decision-making, the collected data are related with the goal and data as the main line. Finally, the relationship between the data and the information itself is analyzed, and the best combination is selected. In order to better play the role of many information sources collected, in the actual process, because the object and environment to be studied are often in a dynamic environment, dynamic scheduling and management of data collection and combination are needed to improve the efficiency of the whole fusion system.

2.   Fusion decision based on statistics

When there is a clear mathematical functional relationship between the decision input and the decision output, the statistical decision-making method is often used, and the best statistical decision-making should be achieved. Bayesian method, regularization theory method and various filtering algorithms constitute the main body of statistical decision analysis. Considering that the basic methods of statistical decision-making are often closely related to the specific decision-making level in the practical application of information fusion, we can further deepen and expand different aspects of statistical decision-making methods. Dynamic statistical decision-making estimation is still the main aspect of statistical decision-making research and application, which is outstanding in content, theoretical method and application. The influence of prior knowledge, accuracy and uncertainty of sensor configuration and the requirement of collaboration are often associated with statistical fusion decision. In addition, the statistical fusion decision-making in the case of multi-target and maneuvering target has also been widely used in practical applications. There are still challenges for improvement of related methods.

3.   Fusion decision based on imprecise reasoning

When the deduction between decision input and output is mainly based on judgment variables, the establishment of decision function is mainly realized by using imprecise reasoning. The reason is that it is difficult to achieve such strict reasoning conditions under normal circumstances. The uncertainty of input data, decision function, processing process and the influence of internal and external noise inevitably lead to the uncertainty of fusion results. The basis of research and evaluation is the expression of uncertainty, which is very important for fusion decision-making, and then the corresponding reasoning method is established based on the expression. At present, there are two kinds of expression methods: the expression based on probability type and the expression based on fuzzy set. The methods of expression and reasoning based on probability type include subjective Bayesian method and evidence theory method. The methods of expression and reasoning based on fuzzy set include fuzzy set and fuzzy logic method, fuzzy integral method and possibility theory method. The reliability of information fusion and the visualization of uncertainty should also be focused on. The reliability of information fusion is to further consider the stability of fusion decision results, that is, the first-order stability of uncertainty.

4.   Intelligent model and fusion decision

For complex information fusion tasks, from input, formation to final decision-making results, the variables involved are multi-level, decision-making tends to be multi-faceted, multi-channel synthesis. It also needs many aspects of history and online accumulated knowledge. It is a complex processing process, which cannot be solved by a single statistical or inaccurate reasoning method, nor can it just rely on the simple stacking of basic methods or general combination of algorithms. We need to use the theoretical method of intelligent model to solve the problem. Using the theory of artificial intelligence and expert system directly can solve some problems, but it needs some basic, standardized and characteristic intelligent processing modes, which affect the structure of the whole fusion system to a certain extent. Distributed processing is more and more widely used. New models such as BN, agent and multi-agent system, ontology are emerging. BN emphasizes the causality and its application in the process of fusion processing, mainly focuses on the distributed processing based on nodes. Agent mainly focuses on the distributed processing based on units or modules. The middle-level structure between the input information and the final decision-making results in the agent can modularize and standardize the processing units needed by the foundation. The multi-agent system intelligence can be realized by some methods, functions, processes, search algorithms or reinforcement learning. Its most direct effect is to simplify the structure of the whole fusion process and facilitate fusion processing. Ontology mainly focuses on distributed processing based on concepts and intermediate processing results, which expresses concepts and intermediate processing results in a clear and standardized form. It simplifies the expression of fusion problem and speeds up the design and implementation of fusion processing.

3.4.3 Fusion Technology Based on Precise Spatiotemporal Information

1. A multi-source sensor management method based on agent

Multi source asynchronous heterogeneous data has the characteristics of numerous sources and large structure differences. Combined with the precise spatiotemporal information provided by Beidou system, we can establish a unified data and communication standard for heterogeneous IOT sensors. An agent-based multi-sensor management method can be formed, providing basic functions such as device management, data access, protocol analysis, and establishing a high-performance system with read–write and compute optimization based on time series data. This method can cooperate with the end of the time series database seamlessly and in real time. The structure of sensor network consists of three parts: computing node, sensor and communication network. The computing node is mainly responsible for the direct management of the controlled sensor (data source), such as data acquisition, working configuration of the sensor, time calibration, range calibration, etc.. The communication network is responsible for the communication between the computing nodes and the communication with the control center. The whole sensor network is a multi-agent system, each computing node is a main agent, and different main agents can coordinate and optimize to build the optimal monitoring network for a certain target. Under this structure, according to different monitoring tasks and targets, the management method of sensor network sub tasks can be realized.

2.   Multi source heterogeneous information fusion processing method

Multi sensor data and multi-source information may come from the same platform or multiple platforms. Each data source has different data types and sensing mechanisms. Data sources cannot keep synchronization, and the perceived target, event or situation may change. Through the establishment of multi-source asynchronous heterogeneous data fusion structure model and data quantitative fusion processing method, multi-source historical and real-time data can be structured to achieve the cleaning, conversion and integration of multi-dimensional data. The information obtained by each sensor is the description of a feature data in the sensor space. Due to the differences of physical characteristics and spatial position of sensors, it is difficult to fuse the information in different description spaces. Before fusion, the information must be mapped into a common reference description space, and then fusion processing must be carried out to get the consistent description of feature data in this space. Therefore, using Beidou precise spatialtemporal information, the sensor data can be unified into the same reference time and space to complete data registration. In the multi-sensor information fusion system, multi-source and uncertain information make the evidence of information source conflict. This conflict is not caused by a single evidence. It may be caused by the error of two evidences, some uncertain reasons, various external disturbances and other factors. When using Dempster combination rule to combine conflict evidence, it will produce a conclusion contrary to intuition, which is the combination of conflict evidence. Therefore, how to achieve the effective fusion of multi-source information in the case of highly conflicting evidence is an urgent problem to be solved. The abnormal data analysis model based on generalized regression neural network can establish evidence theory method based on BP neural network, form object level data fusion processing method based on multi-scale fusion, and achieve multi-source heterogeneous sensor information processing and calculation.

3.   Data compression and nonlinear multi-source information fusion in complex environment

Complex systems will have a large amount of information to processed, but the information transmission is usually affected by bandwidth and energy. While the sampling rate of the sensor is affected, it will also cause information congestion at the information collection end. In order to ensure the measurement accuracy, it is necessary to speed up the information flow when the sampling rate and resolution cannot be reduced. The measures that can be taken in this aspect are usually divided into hardware and software. Hardware, such as increasing the number of channels, improving channel bandwidth and so on. Software can reduce the number of transmission bits by compressing the transmission data, that is, quantization or dimension reduction. Generally speaking, because the hardware measures will not cause data loss, the improvement effect is better, but the operation cost is high. Therefore, it is less in practical application. The software measures can find a balance between the cost and the goal, and are widely used in practical engineering, such as the optimal compression strategy under the constraints of bandwidth and energy. In addition, many fusion estimation algorithms are designed in the framework of Kalman filter. The results of optimal fusion estimation are obtained in linear Gaussian system. In practice, many complex systems are nonlinear coupled systems. For the weak nonlinear system, it can be linearized by Taylor series expansion, and then fusion estimation, but this approximation method is not suitable for the strong nonlinear system. At the same time, for the nonlinear complex system, once the system state or external interference surge, the system model adopted by the fusion filter does not conform to the linear small disturbance assumption. If the linearization estimation method is used, the estimation error will be formed. At present, for nonlinear systems, single channel nonlinear filters such as extended Kalman filter (EKF), unscented Kalman filter (UKF) and particle filter (PF) have been well developed, and many systematic results have been obtained.

4. Effectiveness evaluation criteria of multi-source information fusion

In the multi-sensor system under complex environment, in order to meet the requirements of information diversity, information capacity, information processing speed, accuracy and reliability, it is necessary to evaluate the effectiveness of system information fusion on the basis of modeling and quantification. The effectiveness analysis and judgment of multi-sensor information fusion are mainly reflected in three aspects. The first is the complementarity of information. Information fusion does not mean that the more information, the better. Only the complementary information can reflect the problem from multiple perspectives. Through the fusion processing, the integrity and correctness of the system description environment can be improved, and the uncertainty of the system can be reduced. The second is the redundancy of information. The fusion of redundant information of the same object can reduce the uncertainty caused by measurement noise and improve the accuracy of the system. The third is the effectiveness of the fusion algorithm. For the same fusion information, efficient fusion algorithm will bring more accurate fusion results, that is, the fusion effectiveness is different.

There are many factors that affect the effectiveness of multi-sensor system information fusion, such as fusion algorithm, information selection, correlation between input information and output information, fusion structure and so on. We should not only study and analyze the effectiveness of system information fusion from a qualitative point of view, but also study the quantitative indicators of effectiveness. We should not only consider the selection and correlation of information, but also consider the impact of fusion algorithm on the effectiveness of system information fusion. In view of the possible incompleteness and fuzziness of fusion data, it is necessary to establish the effectiveness evaluation criteria and risk situation evaluation model of multi-source asynchronous heterogeneous data fusion. It is also necessary to establish the subsystem level state estimation method based on dynamic adjustment of false alarm rate, and the system level coordination method based on relaxed membership constraints, so as to timely and effectively evaluate the data fusion effect. We have to establish the input correction state prediction method and real-time data trend analysis method to achieve the visualization of sensor fusion data.

5.   Multi source information fusion system

Multi source information fusion system needs to achieve real-time, multi-dimensional, multi-source, efficient and high-precision online monitoring of various elements, as well as processing, storage, analysis and management, expression evaluation and decision support of monitoring information. In addition, it also needs to make full use of complementary or redundant information of each monitoring sensor in time and space to coordinate and synthesize multi-source information. Data fusion processing system obtains data from various information sources. There is no clear boundary between sensor system, information management system, expression evaluation and assistant decision support system and information fusion system. There is coupling and feedback between them. However, from the system point of view, it can be divided into two levels. The first level is data acquisition and fusion processing, while the second level is information management, expression, state evaluation and auxiliary decision support. The first level is an information processing system that automatically analyzes and synthesizes the observation information of several air and ground multi class, homogeneous /heterogeneous sensors acquired according to time sequence under certain criteria, so as to provide comprehensive, real-time and accurate data for the second level.

Homogeneous/heterogeneous sensors and wide area network environment system are the hardware basis of information fusion. Multi-source data is the processing object of information fusion. Coordination and comprehensive processing are the core of information fusion. The design of data fusion system is distributed structure and hybrid structure. The whole system is composed of two levels of fusion sub centers of a global fusion center. The global fusion center mainly fuses the fused data of each fusion sub center with other relevant information from the outside, presenting a distributed hierarchical fusion structure, while the fusion sub center adopts a centralized hybrid structure. In the global fusion center and each fusion sub center, the information fusion processing is hierarchical. Among them, the fusion sub center is mainly responsible for the information processing of data layer and feature layer, while the global fusion center is mainly responsible for the information processing of feature layer and decision layer.

6.   The application of gas pipe network

Beijing Gas Group Co., Ltd. has formed an intelligent risk assessment and management system for gas pipeline network by comprehensively using Beidou precision service and IOT technology. Facing the intelligent management and control needs of gas pipeline network, and based on the multi-source asynchronous heterogeneous information fusion method of precise spatio-temporal information, the system can provide reference for pipeline defect identification, three-dimensional automatic drawing, fault diagnosis and automatic early warning subsystem.

Based on the Beidou precise service pipeline defect identification subsystem, using the Beidou precise positioning information of the damaged points in the pipeline, combined with the gyroscope, accelerometer, odometer, magnetic compass and data collector and other sensor information, the gas precise model under the dynamic urban environment is established. The research on the precise positioning method of underground pipeline defects is carried out, and the effective early warning decision model is formed.

The pipeline 3D automatic drawing subsystem relies on specific devices to obtain the image data set of the object to be measured in the scene, and accurately corrects the acquisition coordinates through the differential GNSS technology, so as to obtain accurate key data positions such as gas pipelines, appendages and feature points. The system builds the automatic terminal and the corresponding application program to acquire the pipeline data. Computer vision teaches the computer to solve its real spatial attributes, such as shape, orientation and posture, according to the acquired coordinates, reconstruct the real three-dimensional model of the object, perceive the external world through image simulation, and automatically extract the necessary information from multiple images, so as to intuitively present the location of gas pipeline, line patrol route and route in the three-dimensional scene. The location of hidden danger, the location of other ownership pipeline crossing points, etc. are refined to present the real scene.

According to practical statistics, more than 70% of pipeline accidents are caused by external factors such as excavation construction and natural disasters. Aiming at the problem of preventing and detecting the third-party damage accidents, the damage audio library suitable for urban environment is created to solve the problem of feature recognition of geoacoustic detection signal and infrasound leakage signal in complex urban environment. The multi-source warning behavior threat clustering analysis method based on neural network and genetic algorithm is used to analyze and obtain the same kind of behavior from a large number of early warning information. The joint threat recognition mode and early warning technology of multi processing units and the early warning monitoring and command management system are developed to achieve rapid early warning and reduce the false alarm rate and missing alarm rate.

In order to prevent stations, which is the key part of municipal pipe network, from abnormal operation or even paralysis caused by lightning, a small area multi lightning current sensing monitoring network is established, integrated with Beidou precise spatialtemporal information. The lightning monitoring method based on hyperbolic parameter positioning method, precise lightning current positioning method and small area lightning density measurement method are proposed, and the station lightning monitoring system based on Beidou spatialtemporal information is established. The system provides technical support for the lightning protection design of the new station and the transformation of the lightning protection facilities of the old station. Voltage regulator is the core equipment of gas station. According to the characteristics of many factors affecting its fault prediction, a fault diagnosis method of voltage regulator operation based on wavelet analysis and neural network is proposed. A safety early warning device is developed to realize the health monitoring and fault early warning of voltage regulator, change the maintenance mode from regular maintenance to on-demand maintenance, and extend the average maintenance interval.

3.5 Big Data Analysis

3.5.1 Overview of Big Data Analysis

We are in an information era. In such an era, the data existing in the computer can bring people benefits and make people profit from it. At the beginning of the development of computer technology, people use computers or other storage methods to store the data they need. But with the development of the times and the popularity of the Internet in the world, a large number of heterogeneous data need a huge data set to manage and store it. In the analysis of some large data sets with a single association, there will be additional information. These problems make the traditional data processing methods and management methods more and more difficult to apply. Big data came into being.

The concept of big data can be defined as a database with larger acquisition and storage capacity than conventional database tools. Its connotation can be described by four Vs. The four Vs refer to Volume, Variety, Velocity and Value, respectively. Among them, the size of Volume database software is in the order of tens of terabytes to several petabytes. This data will continue to expand with the development of the times and the needs of data management. It must have a certain order of magnitude to be called big data. Variety refers to a wide range of data sources with their own characteristics. It can be monitoring data from sensors, audio and video data, web pages from the network, and various kinds of information in daily life. Velocity usually contains two aspects of information. On the one hand, the frequency of data generation is fast, the update speed is fast, and the growth speed of data volume is fast. On the other hand, the response to data is fast. Data processing needs strong timeliness, and follows the law of one second—the result must be obtained in one second. In the daily use of the Internet and smart phones, Baidu needs to process tens of petabytes of data every day, while Taobao receives tens of millions of transactions every day, and needs to process about 20 TB of data. This requires that the data processing must be fast and efficient. Value contains three meanings. Firstly, in big data, the amount of useful data that needs to be processed is relatively small. For example, when processing video data, the data that may be useful in continuous monitoring images is only a few seconds, which reflects the low value density of big data when processing data. Secondly, the overall value of processing data is high, and when dealing with a problem or an area, the value of processing data is high. It is quite valuable to have a large number of real and reliable data related to the field of research. Finally, big data has a large amount of data, among which the potential valuable data still needs to be mined. It has great potential value.

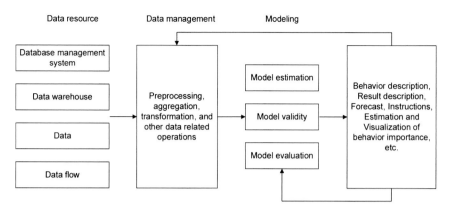

Fig. 3.48 Big data workflow analysis process

Nowadays, big data is no longer a new term, such as in the field of meteorology and biology. When analyzing useful data, large high-end computers are needed to analyze and process big data. At this stage, big data uses large-scale distributed processing technology to improve the efficiency of data processing. In addition, the popularity of cloud computing has changed the way of data storage, computing and access. Since then, the software and hardware environment of big data no longer needs to be built by itself.

Data modeling methods include database, data flow, data set and data warehouse. In the process of data processing, the first step is to integrate, clean and filter the data to ensure the order of magnitude and diversity of the data. The second step is to prepare the data, which is the most energy-consuming step in the process of data analysis. Therefore, we must improve the efficiency of data storage, filtering, transplantation and retrieval before this step to ensure that the follow-up work can be carried out efficiently. Figure 3.48 shows the big data workflow analysis process.

The diversity of data forms makes the analysis and processing of big data more complicated. The data types can be divided into four types as shown in Fig. 3.49. But nowadays, for the data to be processed, it is often not a single one of the four types.

Figure 3.50 shows the classification of data arrival and processing speed, which is the arrival time of data under different classification conditions. The arrival and processing form of data can be continuous, or real-time, or batch. No matter which form, the received data needs to be processed and responded in time.

3.5.2 Common Tools in the Process of Big Data Analysis

1. Hadoop

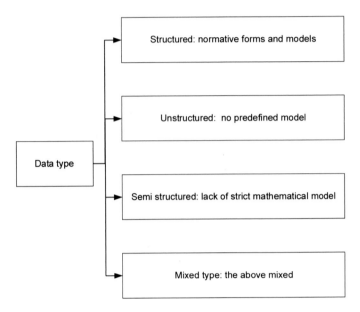

Fig. 3.49 Data type

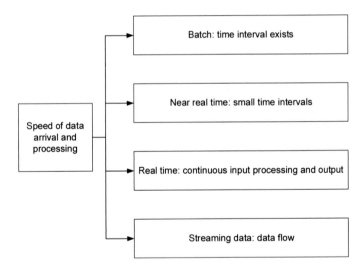

Fig. 3.50 Classification of data arrival and processing speed

Hadoop is the infrastructure of a distributed system. Its advantage is that it allows users to develop distributed programs without having to know too much about the underlying details of the distributed structure. Its operation depends on the community server. Because of its low cost, it can be used by anyone. In the working process, Hadoop speeds up the data processing by working in parallel. It maintains multiple copies of the working data and ensures that these nodes are redistributed when computing elements or storage fails. At the same time, Hadoop can process PB level data. It implements Hadoop distributed file system (HDFS). This distributed file system is attached to low-performance hardware devices, providing high throughput and convenient for users to access application data. It is suitable for large data sets and provides storage for massive data.

2. HPCC

HPCC can be regarded as an excellent substitute for Hadoop. It is more and more prominent in big data analysis. It integrates Thor and Roxie clusters, common middle-ware components, external communication layer, provides client interface for end-user services and system management tools, and supports auxiliary components to store data from external data sources. Figure 3.51 shows the high-level HPCC architecture, which describes the process of big data management of components belonging to HPCC system architecture under mutual cooperation.

Data extraction (THOR) cluster is mainly responsible for processing, trans-forming, linking and indexing a large amount of data. Query (ROXIE) cluster can provide users with high-performance online query function and data warehouse func-tion. Enterprise control language (ECL) combines data performance and algorithm implementation through transparent parallel programming language to operate big data. ECL IDE is an integrated development environment that integrates develop-ment, debugging and testing. Enterprise service platform (ESP) accesses ECL query through HTTP, XML, SOAP and REST, which is a convenient interface.

3. Storm

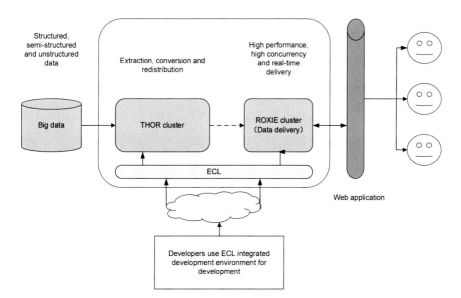

Fig. 3.51 High level HPCC architecture

Storm is an open source distributed fault-tolerant real-time computing system, which is similar to Hadoop programming model. Storm can transmit data through network, process data in real time, and process unlimited data streams reliably. Storm deals with continuous static streams instead of static data. The advantage of storm system is that it provides a simple programming language. Developers only need to pay attention to the logic in the application. The effect is remarkable in dealing with large amount of data. With its continuous development, the amount of data and calculation can be processed is also increasing. The failure of a single node in the operation will not affect the application of the whole system, and ensure that the information will not be lost in the process of processing information.

4.   Apache Drill

Apache Drill is an open source query engine with low latency for Hadoop. It implements Google interactive data analysis system. In this system, large-scale clusters can be built to process PB level data. This data analysis system can reduce the time of data processing to seconds. The features of Apache Drill include: using semi-structured / nested data structure, real-time data analysis and rapid application development, achieving good compatibility with existing SQL environment and Apache Hive.

5.   RapidMiner

RapidMiner is the world-leading data mining solution. Users can drag modeling on the graphical interface of the system without programming. The operation speed is fast. It has more predefined machine learning functions and third-party libraries than other visualization platforms, and can easily access structured, unstructured, big data and other types of data. The system is easy to implement the deployment of data preparation, machine learning and early warning mode.

RapidMiner Cloud can be used as a supplementary computing power. When necessary, analysis models can be deployed in the cloud environment to facilitate users to analyze and model data in any environment.

6.   Pentaho BI

The server of Pentaho BI consists of a BI platform and a library with the ability of final delivery. The server of Pentaho BI makes many functions of BI platform consistent. The content produced by each component is related to the role of each user. It takes process as the center, and the framework is mainly oriented to the solution. The workflow engine is the key controller. The platform is built on the basis of server, engine and component. The components and reports in the platform can be used to analyze the process performance.

3.5.3   Basic Methods of Big Data Analysis

1.   Visualization analysis

Visualization analysis is to show the data that needs to be processed by means of graphics, which is convenient for data display and analysis. It is easy to be understood and received by users. Then, the concept is materialized, and finally the efficient correlation analysis of data is realized. Visualization analysis is a new revolution of data display and human–computer interaction, which combines the aesthetic form and data function. Visual analysis is a kind of analyzer, which is widely used in the association analysis of huge amount of data. With the help of visual data analysis platform and auxiliary manual operation, the process of data analysis and the trend of data link are fully displayed in the form of chart.

Data visualization technology includes data space, data development, data analysis and data visualization. These methods can be divided into geometry based, icon based, image-based, pixel oriented, hierarchical and distributed technologies according to different principles.

2. Data mining algorithm

Data mining is the process of mining hidden information in a large number of data through corresponding algorithms. It is the theoretical core of big data analysis. Different data mining algorithms are based on different data types and formats. The application of data mining algorithms can improve the speed of processing big data. At present, the most influential data mining algorithms are C4.5, k-means, SVM, Apriori, EM, PageRank, Adaboost, KNN, NaiveBayes and CART. Data mining algorithm is a kind of trial and calculation of data mining model. Firstly, the data is analyzed with matching algorithm, and then the corresponding types of patterns are found before modeling.

3. Predictive analysis

Predictive analysis is the core application of big data. Its advantage is that it can transform the relatively difficult prediction problem into a simple description problem, and make the results simple and objective, which is conducive to decision-making. The process of predictive analysis can be summarized as mining the characteristics of the required data from the huge amount of data, modeling the offspring into new data, so as to make a reasonable and objective prediction.

4. Semantic engine

Semantic engine can not only analyze the literal meaning of the content that users input into the search engine, but also analyze the essence of the content. It can comprehensively and accurately grasp the search intention of users, and provide users with better feedback results. The key of reasoning and knowledge storage in semantic engine is knowledge base. When describing a certain domain, ontology can provide a group of concepts, and knowledge base uses these terms to express the facts of the domain. For example, when describing the symptoms of a patient, the knowledge base will describe the specific results of the patient.

5. Data quality and data management

In order to better ensure the authenticity of the analysis results and make the results more valuable, it is necessary to ensure the reliable value and authenticity of the analysis results and obtain high-quality data.

3.5.4 Big Data Processing Flow

1. Collection

Data collection is the cornerstone of the big data industry. In the actual process, because a large number of data are blocked in different software systems, the amount of data is huge, and there are many kinds of data sources. Therefore, the data collection is not easy. In the process of data collection, multiple databases should be used to receive data at the same time for users to process and query.

Data collection is divided into two types, one is collected by web crawler, the other is collected by sensors or other devices. Web crawler is a program that can capture data on the Internet. By grabbing the HTML data of a specific website page, the crawler data can be collected. We can extract the unstructured data from the web page, and store it as a unified local data in a structured way. Using sensors to collect data, we can output the measured information as electrical signals or other forms of information.

2. Import/preprocess

Because there are a lot of incomplete and noisy data in a large number of collected data, inconsistent repetition will appear in some data. There are data that deviate from the expected value. Low-quality data mining results will appear in massive data. It is necessary to import or preprocess the collected data. The main problem encountered in data preprocessing is that the amount of imported data will be very large, sometimes reaching the level of hundred Megabits or Gigabits. Data preprocessing methods include data cleaning, data integration and data transformation. Among them, data cleaning is to remove the noise in the data and irrelevant data. Data integration is to put the data in multiple data sources into a unified data storage area. Data transformation is to transform the original data, so that the data is conducive to the later data mining.

3. Statistics/analysis

Big data analysis will process the original data according to certain analysis ideas, and then conduct artificial analysis on the results. It will summarize different types of information for a certain problem, design statistical schemes, and get more clear and comprehensive conclusions. In the statistical analysis of data, we often face a huge amount of data analysis, and will occupy a lot of system resources.

4. Data mining

Data mining is the most critical step in the big data processing process. It is based on artificial intelligence, pattern recognition, database and machine learning technology.

It uses the corresponding algorithm to calculate the potential patterns of summarized data mining on the basis of existing data to meet the needs of data analysis.

3.5.5 Construction Scheme of Big Data Analysis Platform

1. Requirement description

After running for a period of time, the traditional system cannot make full and effective use of the accumulated massive data, and the existing data processing is too simple. Therefore, the big data analysis platform came into being. In the platform, a large number of data can be managed, analyzed and used to explore the value of potential data. This can solve the problems not found before, make full use of the data, and finally display in a visual way to provide scientific and reasonable decision support.

2. Goal of big data

Nowadays, the big data platform construction of most companies is based on mature open source software, and has carried out optimization and secondary development on these components. The goal of big data platform is not to compare the richness of components and follow up the speed of community technology, but to solve the problems, remove the obstacles, improve the efficiency and increase the income of users on this platform. The goal can be divided into three aspects: to realize the sharing and exchange of data, to realize the collection and storage of big data, and to realize the analysis and decision-making of big data.

The fundamental goal of the construction of big data platform is to improve the horizontal connectivity ability of the internal components of the platform, and to enhance the ability to get through the business process and upstream and downstream links, which can measure the maturity of the platform.

3. Data construction principles

Data construction needs to follow the three principles of data security, data scalability and data flexibility. In order to do access authentication, we need to adopt a high security mechanism, and also need to pay attention to the security performance of the system itself. After the successful construction of the platform, we need to work for a long time, and the scale and requirements of the platform will change constantly. The platform built at the beginning of the design is required to have good scalability. In the application, the platform needs to be integrated with other applications, and in the development, it needs multi-level interfaces to flexibly access other systems.

4. Overall framework of big data

The data analysis platform based on Hadoop technology is generally divided into three parts: data layer, big data acquisition and storage, data analysis and display. Figure 3.52 is a schematic diagram of the data analysis platform.

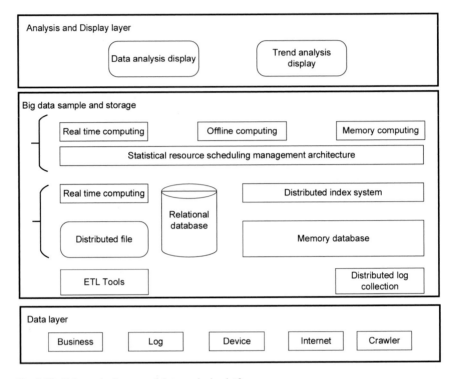

Fig. 3.52 Schematic diagram of data analysis platform

The data layer collects data, limits the target and scope of data collection in the process of collection. Finally, it integrates the collected data to provide favorable and reliable support for the subsequent big data analysis.

Big data acquisition and storage stage is mainly to develop adaptive interface for all kinds of data, so that each type of heterogeneous data can dock with their corresponding system, and also provide data conversion and storage functions.

In order to improve the scalability and fault tolerance of data storage, the HDFS file system with the mainstream big data framework Hadoop, is used to store all kinds of data in a unified way. The data is stored according to the corresponding rules, so as to save a complete set of data files every day and form a data warehouse.

The core business layer of data analysis platform is the final data analysis and display, which can make data reports according to needs. For example, the data analysis system based on Hadoop processes the stored data, generates the report file set transformed by the corresponding algorithm,, and finally displays it in a visual way.

Chapter 4
From Beidou Service Infrastructure Construction to Industry Application

4.1 Precision Service Infrastructure of Beidou Satellite Navigation System

Beidou ground based augmentation system is an important extension of BDS application service. In order to better expand the application of BDS and provide accurate BDS services, various kinds of BDS ground based augmentation networks have been built in different ways all over the country. This chapter focuses on the introduction of two BDS ground based augmentation networks with wide coverage and great influence, namely the national GNSS continuously operating reference stations system constructed by the State Bureau of surveying, mapping and geographic information (Ministry of Natural Resources of China for now) and the Beidou precision service system constructed by the GNSS and Location Association of China (GLAC).

4.1.1 National GNSS Continuously Operating Reference Stations System

The national GNSS continuously operating reference stations (CORS) system is established by the State Bureau of surveying, mapping and geographic information. It is the largest and widest CORS system in China. It can provide real-time sub meter level navigation and positioning services to the public and centimeter level or even millimeter level positioning services to professionals. At present, it has been widely used in land, transportation, water conservancy, agriculture and other fields, and gradually goes deep into the public life.

1. System composition

The national GNSS CORS system started construction in June 2012 and was fully completed in 2017. The system consists of more than 2700 reference stations, including 410 national level stations, more than 2300 provincial level stations

© Publishing House of Electronics Industry 2022
B. Wang et al., *Internet of Things and BDS Application*,
https://doi.org/10.1007/978-981-16-9194-2_4

constructed by provincial surveying and mapping departments, earthquake and mete-orological departments, one national data center and 30 provincial data centers. The system can be compatible with BDS, GPS, GLONASS, Galileo and other satellite navigation system signals, and has the navigation and positioning service capacity covering the whole country, with fast positioning speed, high accuracy and wide range.

2.   System service accuracy

The service system uses a large number of reference stations to receive GNSS signals and send them to the data center in real time through the private network. After calculation, it produces high-precision satellite orbit, clock error and iono-spheric data. All these data are broadcasted to the end users through wired or wire-less networks. Users can effectively reduce the errors in the process of GNSS signal transmission by these data. The positioning accuracy is greatly improved. At present, the system can provide three kinds of precision services.

(1)   Sub meter service: for the public, sub meter service can meet the daily needs of the public, such as lane level navigation.
(2)   Centimeter level service: for professional users, such as surveying and mapping workers.
(3)   Millimeter level service: for special and specific users, this service cannot be obtained in real time and needs professional software for precise post-processing, such as bridge deformation monitoring and settlement monitoring.

Among them, the national data center provides open sub meter level navigation and positioning services for the public, and the provincial data center provides centimeter level and millimeter level services for professional users or special users.

4.1.2   BDS Precision Service System

GLAC is a national industry association in the field of China's satellite navigation and location services. In the process of promoting the civil use of BDS, it has implemented the BDS "hundreds cities, hundreds connections and hundreds applications action" plan. In the process of implementing the plan, through the construction and overall integration of regional and industrial BDS precision service stations, it has formed a national BDS precision service network all over the country, which is now open to the public providing BDS precise location, precise time service and short message communication services in urban gas, heating, power grid, water supply and drainage, intelligent transportation and intelligent pension.

Each service station of the national BDS precision service network has a unique identity code. Through network optimization, it can complete the service coverage of different precision requirements for each service area, and provide 24 h accurate location service. At present, the national BDS precision service system has provided BDS precision service for more than 400 cities in China.

1. Hundreds cities, hundreds connections and hundreds applications action

More than one hundred mature BDS and location-based service application projects have been selected in the "hundreds cities, hundreds connections and hundreds applications action" plan. According to the maturity of the projects, more than one hundred cities have been selected for location-based network interconnection. More than one hundred Beidou and location-based service application projects have been promoted and popularized in each city.

"Hundreds cities" and "Hundreds connections" are applied in more than one hundred cities with better infrastructure and easy implementation in China. According to the relevant national regulations, they realize the standardization of differential signals transmitted by the national BDS precision service system, assign identification codes to the differential signals, unify the format of receiving differential data, and enable users to use a receiving device to achieve cross city and cross regional navigation and positioning. Meanwhile, on the basis of BDS precision service system, the applications of indoor and outdoor seamless navigation are promoted. "Hundreds applications" is to vigorously promote multi industry, multi field and multi-level application within the scope of "Hundreds cities".

2. Industry Application of BDS Precision Service System

(1) Gas industry applications. BDS precision service has been applied in gas pipeline network construction management, inspection, gas leakage detection, gas anticorrosive coating detection, rapid deployment of gas emergency rescue, and monitoring and dispatching of liquefied natural gas tanker.

(2) (2) Application in power industry. BDS precision service has been applied in marketing business application, emergency command and repair, power survey and design, and power timing service.

(3) Application in heating industry. BDS precision service has been applied in heat supply network information collection, network operation inspection, pipeline flaw detection and leakage detection, and emergency rescue.

(4) Water supply and drainage industry application. BDS precision service has been applied in inspection, water supply and drainage precision parts search, rainwater well/sewage outlet/drainage pump collection, flood control and emergency command and dispatch.

(5) Traffic industry applications. BDS precision service has been applied in lane level navigation, driver training and examination, driverless comprehensive evaluation, urban electric vehicle anti-theft management, urban bus intelligent stop management, taxi (online booking) operation management, urban special vehicle precise monitoring management, cross-border port vehicle precise positioning management, traffic infrastructure construction and management, and railway train operation precise control, ship berthing assistance, ship collision avoidance assistance, ship lock management, navigation mark telemetry and remote control, channel dredging.

(6) Application of building (structure) monitoring. BDS precision service has been applied in super high-rise and high-rise building monitoring, bridge

monitoring, long-span building monitoring, deformation monitoring of dangerous buildings, deformation monitoring of historical buildings and heritage buildings.

(7) Safety emergency application. BDS precision service has been applied in regional monitoring of natural geological disasters, regional monitoring of man-made geological disasters, urban priority, remote laser detection of combustible gas by UAV, indoor and outdoor integrated personnel positioning and management of emergency rescue, and command and dispatch of emergency rescue vehicles.

(8) Airport management application. BDS precision service has been applied in civil aviation safety navigation, airport vehicle positioning management and airport personnel positioning and management.

(9) Municipal industry application. BDS precision service has been applied in municipal road public facilities management and street lamp information management.

(10) Application of smart care for the elderly. BDS precision service has been applied in location-based service to assist smart pension, locating the scope and location of the elderly activities, building a safety protection circle for the elderly, combining with intelligent terminal to assist children to remotely understand the health status of the elderly, medical precision service, remote pension for the elderly, and government pension service supervision.

(11) Construction machinery operation guidance monitoring application. BDS precision service has been applied to the guidance and monitoring of piling operation, tower crane operation, excavation operation and ground operation.

4.2 "BDS + IOT" and Industry Application

4.2.1 The Enabling Effect of "BDS + IOT" on the Industry

BDS can directly provide current position and time information of certain users. When it works independently, it can be regarded as a sensor providing space and time information. Therefore, BDS needs to be used as an element to combine with other elements to guide the carrier of land, sea, air and space, assist land exploration and mapping, and plan precise agricultural operations. BDS is like a highly active chemical component (such as hydroxyl ion). Although it only participates in a little, it dominates different combinations.

IOT sensors are closely related to industry applications. Every industry application requires sensors to provide accurate and effective first-hand data, such as soil temperature and humidity in environmental monitoring and agricultural operations, flow and pressure information in gas network, rainfall and displacement in geological disaster monitoring, etc. These data are transmitted to the cloud computing platform responsible for analysis and decision-making through mobile communication network and private network.

Time and location information, or spatiotemporal information in a broad sense, are two essential elements in all human activities. Before the emergence of GNSS, people mainly used long wave timing service, short wave timing service and network timing service to obtain time information, and astronomical navigation, inertial navigation and radio navigation to obtain space information. The emergence and continuous development of GNSS has greatly improved the convenience of production and life in terms of cost and accuracy. BDS, especially satellite based or ground based augmentation signal, can achieve real-time positioning accuracy of meter level, sub meter level and centimeter level, and post-processing accuracy of millimeter level. BDS can also provide nanosecond time service accuracy conveniently and at low cost.

The data provided by IOT sensors is the working information of current time measurement. It is necessary to further analyze and fuse the original measurement information according to its location and time to preprocess the data obtained by various sensors. It also need to integrate the same kind of data, and transform different kinds of data into the same kind by using space–time conversion technology. At this time, it is necessary to fuse the temporal and spatial information provided by BDS with the attribute information provided by IOT sensors, namely "BDS + IOT". The spatiotemporal information represented by BDS and the industry information represented by the IOT can be matched in different combinations, providing basic quantitative perception means for different industry applications. The IOT data with spatiotemporal label can be further integrated by industry applications. As a basic function in different links of the business chain, it provides support for each link of the business chain from two dimensions of deep integration and wide application.

In 2020, among the total output value of GNSS and location based services industry in China is 403.3 billion RMB, the operation service sector accounted for 46.6%, which was the fastest growing sector in all aspects of the industry chain. Moreover, since the end of 2012, BDS-2 system began to officially serve the Asia Pacific region, the proportion of the output value of the upstream links of the industrial chain including basic devices, basic software and basic data and the midstream links of the industrial chain including terminal integration and system integration is decreasing year by year, while the proportion of the downstream operation service links is increasing year by year. The focus of output value has shown a trend of transferring to the downstream, and the industry application is becoming the focus of GNSS and location based service industry. At the end of 2018, the basic system of BDS-3 has been completed, and global services have been provided. The promotion and application of various domestic BDS terminal products has exceeded 40 million sets. The total number of social users of terminal products using BDS compatible chips, including smart phones, is close to 500 million sets. BDS is active in R&D and innovation, with a complete range of domestic products and application systems. The overall performance of the self-developed BDS compatible chip is up to or even better than that of similar international products. BDS signal is supported by the vast majority of global satellite navigation chip solutions. Including mass consumption, smart city, transportation, public safety, disaster reduction and relief, agriculture and fishery, precision machine control, and weather detection, the "BDS + IOT" is used to empower the industry in many fields, such as telecommunications, electric power and financial timing. It also injects new vitality into the global economic and social development.

4.2.2   The Reform Effect of "BDS + IOT" on the Industry

No matter BDS technology or IOT technology, in essence, they are the basic elements of modularization and modularization in the industry application, which can be combined and recursively developed according to the characteristics of the industry. "BDS + IOT" is not only an independent production mode in industry application, but also an open technology to create economic structure and function through the continuous development of satellite navigation technology and IOT in recent years. This modular technology cannot only connect a series of loose technologies together to form a solid basic unit, but also become a standardized component of industry application with the passage of time and the deepening of application. For example, in the urban gas industry, BDS precision service and IOT information fusion technology has penetrated into all aspects of the business chain. The welding construction of underground pipe network, disaster prevention and mitigation, leakage detection, early warning control and intelligent decision-making all need precise time–space information and business information fusion, so as to form an overall solution covering "construction, prevention, inspection, control and intelligence". In essence, this change is brought about by the deep integration of universal accurate time–space information and professional sensor business information, which is the transformative effect of "BDS + IOT" on industry applications.

Technology has hierarchical structure and recursion. Holistic technology is equivalent to trunk. The main technology integration is the secondary branch. The basic technology elements are smaller branches. In this structure, there are some self similar components, that is, technology is constructed by different levels of modular technology. For industry applications, all technologies in "BDS + IOT" are purposefully integrated to solve some problems, and modular technologies are gathered together to achieve the common goal. However, with the continuous development of technology and the deepening of understanding of industry application, the original industry pain points are solved one by one, and new industry pain points will continue to emerge. Therefore, "BDS + IOT" is not a once and for all solution, but to face more problems, to constantly carry out internal replacement and structural deepening. Internal replacement refers to replacing an obstructed part with a better part (sub technology). Structure Deepening refers to finding a better part or adding a new one.

Internal replacement refers to the enabling effect of "BDS + IOT" on the industry mentioned in Sect. 4.2.1, while structural deepening refers to the reform of the industry by "BDS + IOT". The development of the industry brought by new technology is not only to replace some traditional technologies, but to improve all components of the industry at a certain level at the same time. This kind of improvement is not only on the technical level, but also on the organizational management level.

The arrival of a new technology will lead to the reshaping of production and price, as well as the expansionary adjustment and even change of economic model. This change means that the existing industries adapt to the new technology, choose the content they need, and combine some components in the new technology and some components in the new field to create a new secondary industry.

We still take the application of "BDS + IOT" in urban gas industry as an example to illustrate the industry reform and the creation of new industries. In recent years, virtual reality (VR) and augmented reality (AR) technology have been favored by the capital, but its industrialization and industrial application is relatively limited, because it cannot meet the precision needs of industrial applications. Beijing Gas Group, on the other hand, combines BDS with VR and AR. Firstly, based on augmented reality and mobile GIS technology, the underground crisscross pipe network is accurately marked on the three-dimensional map. Then the centimeter level BDS positioning module and attitude measurement unit are installed on the VR glasses to accurately obtain the current position and attitude of the wearer. In this way, after wearing glasses, the pipeline inspectors can intuitively see all kinds of information of the underground pipe network, and can see the accurate relative position between themselves and each pipeline, so as to more effectively carry out leakage detection and other work. The "BDS + IOT" technology is used for the visualized management of gas pipeline network asset data, so that the communication, discussion and decision-making in all aspects of design, construction, operation and management are carried out in a visualized state, which provides an environment for decision-makers to analyze problems, establish models, simulate decisions and formulate schemes, and realizes high-quality and high-level intelligent decision-making.

"BDS + IOT" changes the industry application, in essence, because the deep understanding of one technology can be used to the deep understanding of another technology. This innovative application is not only to constantly find or combine new solutions in the existing technology and practice, but also to effectively combine new technologies and new functions in the process of practice. When a new technology enters an industry application as a new element, it will act as a new node. This node may replace the old technology and its related combination, and the corresponding production and service mode will be readjusted to adapt to the changes brought by the new technology. At the same time, the cost and price will also make corresponding changes, so that the new technology is constantly improving. Therefore, the structure of industry application will not only adapt to the change of technology, but also restructure with the change of technology. Technology creates the structure of industry application, and industry application also regulates new technology.

Chapter 5
"BDS + IOT" and Industry Empowerment

5.1 Application in Transportation Industry

5.1.1 Introduction

Transportation is the economic lifeline of a country. All kinds of production activities and materials needed by people's daily life need to be guaranteed by transportation. Highway automobile transportation is the most common and main means of transportation. With the development of technology, automobile plays a more and more important role in daily production and life. The effective command, coordinated control and management of all kinds of vehicles is an important problem faced by transportation and safety management departments. Statistics show that in recent years, the direct and indirect economic losses caused by road congestion in many countries, including China, are astonishing. In order to meet the needs of improving transportation efficiency and security, countries have carried out the research of vehicle navigation and positioning technology based on GNSS. The intelligent transportation system based on GNSS is a public management platform developed by governments.

In addition to all kinds of material transportation required by the general national production and life, the material transportation and public passenger transport services that need to be monitored and managed, which are related to the safety of people's lives and property, are the objects of monitoring and management in the traffic management department in recent years. In addition, with the rapid development of economy, driven by economic interests, some enterprises or individuals often violate the regulations on transportation management of inflammable and explosive dangerous goods, and transport dangerous goods at inappropriate time and place. All kinds of accidents happen from time to time, which has brought many adverse effects on social security and economic health. Therefore, it is an important task to monitor and manage the transportation process of various flammable and explosive dangerous goods.

© Publishing House of Electronics Industry 2022
B. Wang et al., *Internet of Things and BDS Application*,
https://doi.org/10.1007/978-981-16-9194-2_5

At present, the country has made it clear that for the important industries related to national economy and public safety, we must gradually transition to the service system of BDS compatible with other satellite navigation systems. It is an inevitable development trend to use Chinese independent BDS to achieve the monitoring and management of passenger transport, dangerous goods transport and other key transport processes, which will benefit the development of the national economy and the safety of people's lives and property.

5.1.2 Application Scheme

The application of GNSS in the field of transportation generally adopts the mode of "positioning terminal + service platform", namely "vehicle intelligent GNSS terminal" and "location based service and public information platform". The overall framework of the application of GNSS in the field of transportation is shown in Fig. 5.1.

"Location based services and public information platform" is based on advanced technologies such as GNSS, wireless communication and cloud computing, and adopts cloud architecture system to provide cloud platform services with high reliability, strong scalability, high scalability and openness. It can be deployed in various ways such as public cloud, private cloud or hybrid cloud according to specific objects, as shown in Fig. 5.2. This multi-level cloud service computing environment can

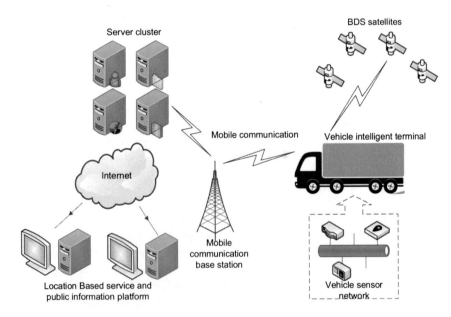

Fig. 5.1 Application framework of GNSS in transportation field

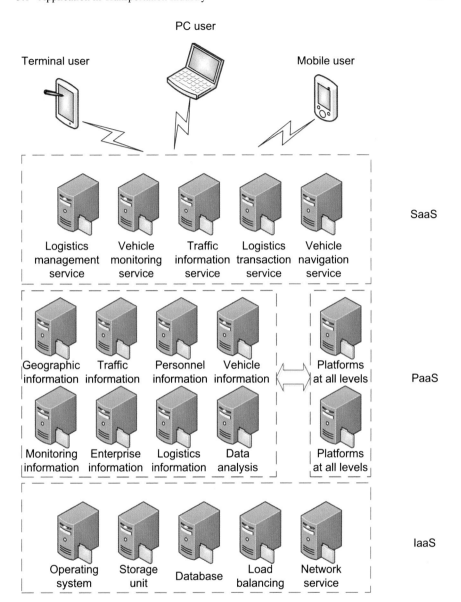

Fig. 5.2 Location based services and the overall framework of public information platform

reduce the difficulty of system development and application, enhance system avail-ability, reduce the overall investment, and reduce the system construction cost. Loca-tion based service platform provides basic services of PaaS (Platform as a Service) level general logistics information business. Its goal is to meet the business needs of manufacturing enterprises, trading enterprises, logistics enterprises, vehicle users, group customers and other secondary developers. Developers can develop SaaS (Software as a Service) system of logistics industry on the basis of basic cloud platform, such as logistics management system, vehicle monitoring system, etc. Location based service platform mainly includes logistics location data cloud storage system, logistics geographic information cloud service system, logistics vehicle navigation cloud monitoring system and logistics enterprise application cloud service system.

PaaS service provided by location based service and public information platform, aiming at the construction of business application system to meet the needs of users in the transportation industry, provides users with transportation e-government, transportation e-commerce, vehicle dynamic monitoring, vehicle location service, safety rescue, information exchange analysis, auxiliary decision-making and other services of cloud computing experience. It improves the efficiency of government and enterprise management, reduce operating costs, and realize intelligent logistics management by means of information technology. Through various basic cloud services such as vehicle monitoring, information transmission, early warning and alarm, safety rescue, information statistics, map browsing and so on, the public information platform of intelligent transportation is realized on the platform. The SaaS service system that can provide business support for government, fleet and other users is built. The platform includes transportation enterprise management, transportation vehicle management, transportation goods matching, user integrity certification, transportation outsourcing service, transportation SaaS service and other functions. It can realize the government supervision and service on the transportation industry, the information exchange and business outsourcing between upstream manufacturing enterprises and downstream business flow enterprises. It can also help transportation enterprises strengthen business management and get through the business process. The operation efficiency of transport vehicles can be improved. The no-load rate can be reduced. The whole visual management of people, vehicles, roads and goods can make the transport activities more efficient and intelligent. Through the data exchange function of the platform, the interconnection and information sharing of all links of transportation can be realized, which is conducive to the realization of cross regional transportation linkage. Public information platform mainly includes transportation public information portal, e-government platform, e-commerce platform and transportation data exchange platform.

"Vehicle intelligent GNSS terminal" communicates with the service platform according to the standard interface protocol. Integrating GNSS, mobile communication, dash cam, vehicle condition monitoring, electronic map, multimedia, smart card/biometric and other modules, it can realize multi-mode satellite navigation and positioning, vehicle illegal monitoring, vehicle driving safety monitoring and warning, task-based navigation, real-time traffic broadcast, vehicle command and dispatch, remote cargo stowage, fleet capacity report, electronic waybill, audio-visual entertainment, etc. As the carrier of vehicle information collection, communication and release, the terminal interacts with the platform online in real time, reflects the service of cloud platform and business system through the terminal function, and communicates with the platform through standard protocol.

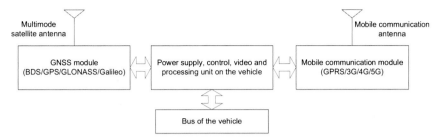

Fig. 5.3 Principle of GNSS vehicle intelligent terminal scheme

The development of vehicle intelligent GNSS terminal is divided into terminal hardware and embedded software. Figure 5.3 shows the principle of the intelligent terminal scheme based on GNSS multi-mode navigation module. The dual-mode navigation module receives GNSS signals and obtains real-time vehicle position, time, speed and other information through calculation and processing. The information required for monitoring is obtained through data processing of on-board monitoring and processing unit. Through two-way data communication with monitoring center through mobile communication network, it can realize alarm, positioning, information collection, human–computer interaction and other functions. It can carry out real-time remote vehicle dispatch, monitoring and management.

The functions of each part of the scheme are as follows:

(1) Satellite navigation module: receives GNSS signal, calculates real-time position, time, speed and other information of vehicle, and sends it to on-board monitoring and processing unit.

(2) On board monitoring and processing unit: runs the control and processing program of monitoring terminal for data processing and peripheral equipment control, mainly including alarm, positioning information and vehicle status display and acquisition, data storage, communication protocol processing and related interfaces.

(3) Mobile communication module: responsible for communication between vehicle monitoring terminal and control center.

5.1.3 Application Functions

Location based service and public information platform can adopt cloud computing, 3S (GNSS, GIS, Remote Sensing) and other technologies access the service request of mobile communication network and Internet. They provide basic services such as map download, location service, terminal communication, terminal control, image monitoring, alarm and early warning, vehicle scheduling, statistical analysis, and support the service docking request of terminal system manufacturers and the secondary development demand of system developers.

Location based services and public information platform are focused on the transportation industry information requirements. They are constructed to meet the needs of government and enterprise application system. They provide different users with cloud computing based transportation e-government, e-commerce, vehicle dynamic monitoring, location-based services, security assistance, information exchange, analysis, decision-making and other services. Information is as a means to enhance government and enterprise management efficiency, reduce operating costs, achieve intelligent transportation management.

Vehicle intelligent GNSS terminal is composed of embedded processor (CPU), read-only memory (ROM), random access memory (RAM), GNSS navigation module, vehicle driving record module, mobile communication module, intelligent card reading and writing module, multimedia service module and other modules. It also provides interface of fuel consumption sensor, temperature sensor and vehicle can bus, in order to achieve multi-mode GNSS, vehicle illegal monitoring, vehicle driving safety monitoring and warning, task-based navigation, real-time traffic broadcast, vehicle command and scheduling, remote cargo loading, fleet capacity reporting, electronic waybill, video entertainment, and with cargo loading, command and scheduling, video entertainment and other functions.

Based on the current situation of users, the system conforms to the development direction of the transportation industry. The system closely meets the needs of multiple users, so as to promote the comprehensive development of all aspects of transportation, as well as the cross industry and cross regional communication and cooperation of relevant departments. It improves the transportation service capacity and efficiency, strengthens the supervision ability of government management departments on the transportation market as the main line, and takes BDS location service as the main line. The multi-party data acquisition, integration and sharing is the core. Information security is the basis. The system provides safe, reliable, effective and real-time information services for industry authorities, transportation enterprises and the public. This fully reflects the development and utilization of information resources among government, enterprises, goods and materials, promotes the establishment of collaborative operation mechanism and strategic cooperation relationship among enterprise groups, and provides support for the government. At the same time, it can provide a variety of transportation information value-added services. In addition, the system also starts from each link of the transportation chain, focuses on the development trend of information service of freight logistics nodes such as freight hub and logistics park, improves the information level of fleet, freight station and logistics park, and carries out standardized management. The BDS can fully play the positive role in the process of freight transportation.

The application of GNSS in the transportation industry can change the situation that the government and enterprise safety supervision are not in place in the current road transportation industry, especially the lack of necessary supervision and information collection means for transport vehicles and drivers. The relative scattered information of transportation is not easy to supervise, and the dynamic connection between supervision and punishment has not yet formed, resulting in the basic credit assessment information of employees is incomplete. By using BDS, we can solve the disadvantageous situation that vehicle location information service mainly relies on foreign GNSS. The promotion of transport vehicle monitoring and supervision platform in the road transport industry can further improve the supervision system function of road transport management institutions at all levels, form the platform level by level assessment management mode, and ensure the long-term operation mechanism of the system. In addition, while meeting the regulatory needs of the government on the transport industry, the transport public information service platform can provide e-commerce platform for production enterprises and logistics enterprises, provide authoritative and reliable transport transaction platform for both sides, and realize the outsourcing and management of transport business.

5.2 Application in Smart Logistics Industry

5.2.1 Introduction

Logistics is a comprehensive industry involving many departments and industries, which contains rich content and involves a wide range of fields. It cannot be simply summarized as transportation. Logistics activities run through manufacturing, commerce, warehousing and transportation, including manufacturing enterprises, agricultural processing and trade enterprises, commercial sales enterprises, highway transportation enterprises, shipping and transportation enterprises, foreign trade enterprises, postal distribution enterprises, financial and insurance enterprises, etc. At the same time, many government departments are involved, mainly including various transportation management departments, industrial and commercial departments, tax, customs, inspection and quarantine, agriculture and other administrative departments. This requires that the logistics and transportation management system based on GNSS should be a medium platform for information exchange. It must have the ability to integrate the information resources of all aspects of logistics activities. It must collect the information distributed in government departments, freight and logistics enterprises, freight terminals, logistics parks, manufacturing factories, trade enterprises, commercial marketing enterprises, and sales enterprises. The information is integrated on a unified platform. User data are fully mined, processed and utilized to become an important information carrier to carry out logistics management and service. Fully exchange and sharing of information is ensured among the government, enterprises, customers and stations. All parties involved are linked up to coordinate and cooperate with transportation management and production activities. The logistics management is further optimized to allocate resources. The main highway hub is fully used in the whole logistics system. In addition, the system can provide a variety of information services for all parties, strengthen the development and utilization of logistics information resources. The system provides reliable, effective and real-time logistics information services for competent departments of transportation industry, transportation and logistics enterprises, logistics stations and the public.

It ensures the full sharing of logistics information resources and provide guidance for industry supervision and management, transportation and logistics management, production and service. Through the construction of the system, it will also improve the application of advanced information technology, such as cloud computing and geographic information system, and comprehensively promote the development of logistics information modernization of user units.

Many functions of intelligent logistics information platform need to be connected through time service, navigation, positioning and tracking services. Therefore, navigation technology including BDS, IOT technology, mobile location information service, geographic information system and wireless communication technology, need to be applied to make the whole process of supply chain logistics transparent and traceable. It conducts comprehensive control and standardized management of operation, and ensure the smooth operation. The accident rate and cargo damage rate are reduced.

The intelligent logistics information platform can obtain the time, location and loading information of various resources in real time through various navigation and positioning technologies. The platform integrates various logistics resources to effectively schedule more social logistics resources. Intensive utilization improves the development level and growth space of regional logistics. According to the services provided by all kinds of location information and geographic information system, the intelligent logistics information platform can provide customers with tailor-made professional, meticulous and personalized supply chain logistics services, and improve the logistics service marketing ability.

According to the requirements of the logistics ecosystem, the intelligent logistics information platform needs to establish an information collaboration mechanism that supports real-time information connectivity and workflow. All node enterprises, logistics enterprises, customers and platform employees in the supply chain can operate in a unified and collaborative manner. We should not only integrate the logistics service information of government departments, but also make the railway, highway, waterway, aviation, postal and other information effectively cooperate on the intelligent logistics information platform. Therefore, navigation, positioning, tracking and other related technologies are particularly important. Using these technologies, logistics enterprises and enterprise logistics can carry out integrated logistics services based on supply chain on the intelligent logistics information platform. It makes the whole process of logistics transparent and visualized, implements intelligent management of the whole life cycle of goods, and further improves the socialization level of logistics services.

The whole process visualization of logistics is the data basis of smart logistics. By combining BDS with other logistics information platforms such as identity recognition, mobile communication, cloud computing and intelligent transportation system, we can expand the coverage of logistics information network and promote the large-scale development of professional logistics information service industry. Smart logistics information platform applies these technologies to automatic identification, location service, information exchange, visualization service, intelligent transportation, logistics management and other aspects. This can influence and drive the development of a number of professional logistics information service enterprises through the application of information technology, and drive the service innovation of supply chain finance through informatization.

5.2.2 Key Technologies

With the development of economic globalization and the continuous expansion of the scale of logistics enterprises, there are more and more logistics links. The amount of logistics information increases rapidly. The demand processing process is more and more complex, and the requirements for the distribution system are also higher and higher. However, the traditional logistics distribution system has low level of informatization and visualization. It has poor ability of analysis and decision support for multi-source and massive data. In addition, the decision-making based on the traditional distribution model is too idealistic and does not fully consider the changes of multiple factors in practice, which is far from the actual application. Therefore, it is very necessary to improve the scientific visualization and information level of distribution system decision-making. BDS provides a means to solve the above problems. It is of great significance to apply BDS spatiotemporal information to all aspects of logistics distribution management and build a comprehensive intelligent logistics system to improve the efficiency of logistics distribution.

Under the development system of intelligent logistics, BDS is no longer regarded as an isolated technology, but as a collection of functions integrated into enterprise development decisions. Logistics management is the management of commodity information. The intelligent terminal obtains the attribute and spatial information of commodities through RFID, laser scanning, infrared identification, navigation and positioning technology. Then transmits it to the data center through the network, and stores and manages the data, so as to provide the basis for monitoring, planning and decision-making for order management, optimal path, warehouse location, goods tracking, vehicle scheduling and value-added services.

Combined with BDS, it can realize fast navigation, positioning and tracking, obtain real-time space and attribute information of vehicles and goods in logistics, and carry out remote and visual management of various elements in logistics, which is very critical for efficient management of modern logistics. Figure 5.4 shows the typical working principle diagram of BDS and management system in intelligent logistics.

In the intelligent logistics system, BDS can not only provide logistics distribution and dynamic scheduling functions, but also provide cargo tracking, vehicle priority, route priority, emergency rescue, reservation service and other functions. BDS can use terminal device to monitor the status of vehicles and goods in real time, which is of great significance to realize real-time deployment. BDS can help people, vehicles, warehouses, routes and areas to make the most suitable logistics plan, and even provide better support for business decision-making such as business division and

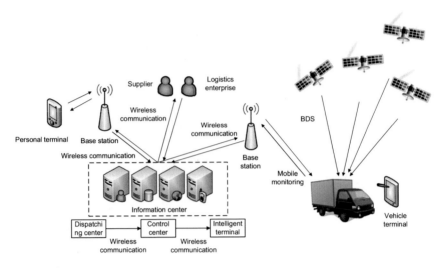

Fig. 5.4 Typical working principle diagram of BDS and management system in smart Logistics

location through the analysis of customer coverage, market saturation, competition and chained network optimization. Combined with navigation electronic map data and real-time traffic information, it can improve the efficiency of vehicle distribution and further reduce the cost of logistics distribution. At the same time, the map and building data provided by GIS can be used to improve the logistics analysis technology. The basic geographic data of GIS provides accurate and detailed information support for logistics distribution, including the location relationship between distribution area, distribution center and customers, and the route planning information from distribution center to customer location. In addition, BDS integrates the latest technologies of relational database management, efficient graphic algorithm, interpolation, zoning and network analysis, and effectively applies the spatial analysis functions of GIS (such as best path analysis, address analysis, buffer analysis, etc.) to the analysis of logistics, which can improve the overall technical level of logistics industry. Combined with GIS, it will further strengthen the integration with ERP, CRM and other systems within the enterprise. The logistics information will also provide the overall decision-making for the enterprise through the navigation and management system.

The leap from modern logistics to intelligent logistics puts forward higher requirements for spatiotemporal information, and also puts forward new challenges for the application of BDS. The smart logistics based on cloud computing urgently needs to realize the combination of cloud computing and Beidou spatiotemporal information. It accelerates the intelligent development of geographic information, and builds a smart logistics system. The concept of cloud navigation arises at the historic moment.

The so-called cloud navigation is to use the various characteristics of cloud computing to support the elements of navigation and geographic information,

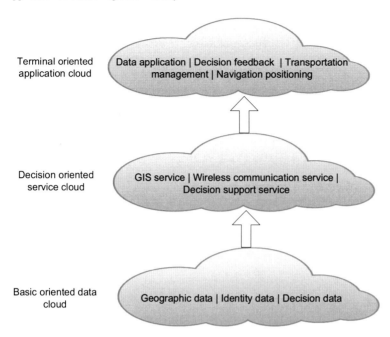

Terminal oriented
application cloud

Data application | Decision feedback | Transportation management | Navigation positioning

Decision oriented
service cloud

GIS service | Wireless communication service | Decision support service

Basic oriented data
cloud

Geographic data | Identity data | Decision data

Fig. 5.5   Cloud navigation platform architecture

including storage, processing, analysis and modeling, so as to change the traditional application methods and modes of navigation, positioning and tracking of logistics users. It uses the navigation and geographic information resources in a more friendly way with low cost and high efficiency. Cloud navigation can provide cloud map slicing service, by uploading cache map slicing to cloud data center, to improve the efficiency of data access and save the cost of data use. Cloud navigation can realize data and information transmission through the web, to solve the problem of efficient data management and real-time technology update of enterprises. Cloud navigation can configure software services according to the needs of enterprises to build a multi environment seamless navigation system. The introduction of cloud navigation into the logistics industry has built a high-end logistics service platform, effectively promoted the development of spatial information application industries such as location-based services, logistics transportation and various value-added services, and helped the formation of intelligent industry chain. Figure 5.5 shows the cloud navigation platform architecture.

5.2.3   Application Scheme

The most extensive application of BDS in the logistics industry is in the construction process of dispatch and distribution logistics system. Distribution refers to the economic behavior of resource allocation according to the needs of users by means of modern delivery. Distribution is a terminal link in the logistics system, which refers to the process of assembling goods in a reasonable form in the distribution center according to the order requirements of retailers, and then delivering them to retailers in the most reasonable way. The distribution cost often accounts for more than half of the whole logistics cost, thus, improving the distribution efficiency has great economic benefits.

A large number of logistics distribution applications mainly adopt the centralized distribution network structure. The main characteristics of the centralized distribution network are low management cost, low safety stock and long lead time of users. However, the distance between distribution center and users is relatively far, and the outward transportation cost (transportation cost from distribution center to users) is relatively high. However, with the continuous increase of retail online stores, the distribution scale is becoming larger and larger, especially for urban and rural distribution. It has the characteristics of small batch, multi batch and scattered delivery points, which puts forward higher requirements for the distribution capacity of distribution center. Therefore, in the process of commercial distribution and transportation, there will inevitably be a series of problems, mainly in the following aspects.

(1)   Unable to effectively supervise the business operation of distribution vehicles. After the distribution vehicles drive out of the distribution center, it is impossible to monitor the driving status, safety status and driving path of the vehicles. Therefore, the distribution vehicles will deviate from the route without authorization, or stay at a customer place abnormally.

(2)   It is impossible to adjust the route in the process of logistics distribution in real time according to the actual situation. The overall utilization rate of distribution vehicles is low. For example, if a company can meet the delivery demand of more than 800 pieces/day in the peak sales season. When the delivery demand is only more than 300 pieces/day in the off-season, it does not need the same number of delivery vehicles to complete the task.

(3)   Failed to deliver goods to customers. Generally speaking, the requirements of large customers such as supermarkets can be met. But in the actual distribution process, due to the influence of traffic control, customer geographical location and other objective factors, it cannot fully meet the requirements of small and medium-sized customers.

(4)   Delivery personnel violations occur from time to time. Without any monitoring, it is difficult for distribution center to manage the distribution personnel in each work place. Some distribution personnel take advantage of this management blind spot, fail to complete the work in strict accordance with the delivery process, and arbitrarily "cut off" the flow, so that some customers' orders cannot be fulfilled in time, causing damage to the interests of both the supply company and customers.

The above problems can be attributed to the lack of vehicle positioning and vehicle route tracking. At the same time, the distribution center is lack of navigation tools to guide each distribution personnel on the road, resulting in the vehicle cannot be flexible scheduling, and rely too much on the experience of distribution personnel, increasing the cost of distribution and staff training costs. In particular, it lacks process standard system to form responsibility traceability. And the communication between retail outlets and distribution centers is not close, resulting in regulatory loopholes.

At present, most logistics distribution centers in China have realized the optimization of traditional routes and vehicle loading scheduling in the central control system using BDS. The map data, positioning data, retail investor data, distribution vehicle personnel data and order data are obtained from BDS, distribution system and telephone ordering system. Based on the clustering optimization algorithm, the scientific calculation is carried out to form the optimized route, variety and formula plan and other information that can be used to complete the distribution task, which is solved by the intelligent distribution management system.

The flow of modern logistics distribution is shown in Fig. 5.6. Retail customers send order information to customer service center through phone, website and other channels. Then customer service center sends the approved order data to distribution center. The distribution center generates distribution lines and corresponding time plan, and specifies corresponding distribution vehicles for each distribution line. Then the strategy information is sent to the warehouse. The warehouse sorts the goods according to the information. The assigned vehicles deliver the sorted goods to the retailers.

Fig. 5.6 The process of modern logistics distribution

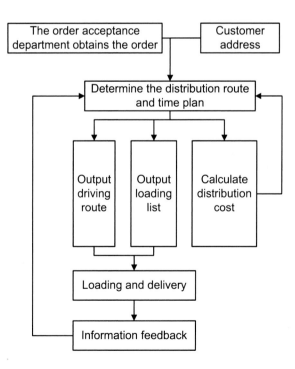

The basic idea of logistics management platform is to introduce GIS and BDS into goods distribution business to realize the distribution process shown in Fig. 5.6. The overall design idea of the system is as follows.

(1)   In terms of application, under the principle of giving priority to the lowest cost, the full load of vehicles is considered, and the workload of distribution personnel is weighed to plan the distribution route. Through real-time positioning of BDS, the distribution vehicles, distribution results and other information are monitored. The monitoring data is stored in the data center database for other systems to query and use.

(2)   In terms of data demand, geospatial data (road data and retail location data) and logistics information data (vehicle data and order data) are collected and integrated with order information.

(3)   In terms of technology, GIS, BDS and web technologies are introduced to combine the customers' spatial location and road network information, and change the original table format management into intuitive spatial management and network management, so as to realize real-time navigation, monitoring and networking of goods distribution.

According to the above ideas, the logistics management platform should include retail geographic information database, intelligent generation of distribution lines, BDS vehicle monitoring, comprehensive geographic information analysis and other functional modules. This structure is based on the management and application with the target customers (retail customers) as the core. It provides a systematic solution from the management, monitoring and scheduling of distribution vehicles with the target customers to the management, assessment, analysis and statistics of retail customers. The platform is also linked with the business and monopoly system to express the key business information on the electronic map in real time and dynamically, providing managers with an intuitive result of sales and operation status, forming a GIS application system suitable for the industry.

In the design of the overall structure of the system, the use of traditional client/server (C/S) mode and browser/server (B/S) mode can help to decompose transactions, improve the processing speed of server and client, and ensure the security and sharing of data. However, in the process of distribution, it is quite frequent to determine the location of retail customers outdoors, and a large number of maps and retail customers information query and statistics are involved in outdoor tracing. At the same time, it will inevitably produce a large number of updated data. These cannot be satisfied in the previous C/S and B/S system framework. Therefore, mobile/server (M/S) mode is introduced in the design of the overall framework of the system. A system framework combining C/S, B/S and M/S is established. In the framework of C/S, B/S and M/S, the internal functions of the system are modularized to ensure that after the development of each functional module of the system, different modules can be selected for combination according to the business needs. Different business processes can be satisfied by setting the functions of the modules and the relationship between the modules, so as to improve the reliability, heritability, reliability, maintainability and expandability of the software. This truly implements the principle of

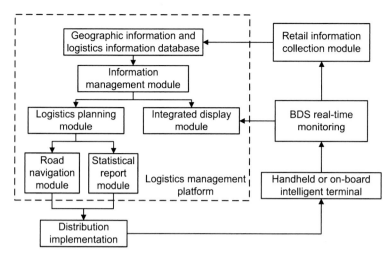

Fig. 5.7 System frame structure and function decomposition diagram

overall planning and distributed implementation in the construction of the system, so that the system can be constructed and improved step by step and in batches according to the characteristics of system engineering. Figure 5.7 shows the system framework and function decomposition diagram.

According to the business requirements, combined with the integrated framework design of C/S, B/S and M/S, the intelligent logistics integrated information management platform is divided into the following main subsystems.

(1) Intelligent generation system of distribution route. The system is established in the C/S mode, mainly serving the various departments of the distribution center of the supply company and its subordinate teams and other level business units. Its main function is based on the order data sent by the business system, which comprehensively analyzes the road network traffic information data, vehicle information data, retail information data, etc. On the basis of meeting the requirements of logistics task and delivery time, it provides optimized logistics distribution scheme, so as to arrange logistics vehicles and personnel to save costs. The input data of the system comes from the business system. The output data is the sorting and distribution strategy data provided to the distribution and sorting system and the road navigation route data and delivery order data provided to the fleet scheduling. The system should have the functions of data editing, data management, path network analysis and assistant decision-making. Using GIS platform, through the internal LAN to access the spatial data and business data stored in the database, it can provide information management, planning and scheduling, information query, comprehensive display, report statistics and other functions for all levels of personnel in the distribution center.

(2)   The handheld GIS/BDS and vehicle navigation system based on M/S mode are
      adopted. The handheld GIS/BDS terminal based on M/S mode aims at outdoor
      work site, extends the indoor information system, and provides on-site appli-
      cation services for on-site staff, such as map browsing, retail positioning, point
      drawing on electronic map, on-site business data collection, correction and
      batch import of logistics management platform database, BDS real-time navi-
      gation and other functions. Based on smart phones or pad, the retail batch
      positioning system integrates BDS, electronic map and management software,
      providing a new technical means for collecting and updating retail information.
      BDS differential positioning technology can provide sub meter accuracy posi-
      tioning information in a short time. At the same time, using the management
      software of mobile GIS device, we can collect detailed attribute data while
      obtaining positioning information, and display the retail location in real time
      on the electronic map. Through the system, operators can complete the work
      of retail positioning, point tracing, correcting drift and batch importing into
      the logistics management platform database. The vehicle navigation system
      is composed of tablet computer and BDS receiver, communication module,
      management software and imported navigation route data, which provides
      navigation tracking function for distribution vehicles. The BDS receiver of the
      system is responsible for receiving the coordinates of the geographical location
      of the distribution vehicles, and using the management software to accurately
      display the location and operation status of the vehicles on the imported distri-
      bution route map. This is to realize the real-time navigation and monitoring
      of the distribution vehicles, and get rid of the excessive dependence on the
      dispatcher's and the driver's experience. The navigation tracking of delivery
      vehicles improves the operation efficiency of delivery vehicles, reduces vehicle
      management costs and resists risks.

(3)   WebGIS command center system based on B/S mode. The system uses GIS
      platform and data server. The main service object is the management personnel
      at all levels of the supply company and business query personnel, so that they
      can also express the information and sales data of each dealer on the electronic
      map in the form of thematic map in real time and dynamically through the
      browser without installing the application system client to query business data.
      The location information of logistics vehicles is sent back to the command
      center through the communication module by the vehicle BDS navigation
      device, which matches with the electronic map on the central system. The
      command center can also intuitively grasp the dynamic location information
      of logistics vehicles, so as to realize the monitoring and scheduling of mobile
      terminals.

5.3 Application of Power Timing Industry

5.3.1 Introduction

Satellite navigation timing service system is one of the key national infrastructures. Precision time is a basic physical parameter in scientific research, scientific experiment and engineering technology. It provides an essential time reference for the measurement and quantitative research of all dynamic systems and time series processes. Precision timing service has a wide and important application in the industrial fields such as power, communication, control and national defense. BDS-1 uses 2.4 GHz frequency. Because the 2.4 GHz signal is easily interfered by WiFi, microwave and other adjacent frequency points, it has a great impact on the antenna location, installation and BDS timing service performance in the construction of BDS-1 time service project. It is unable to ensure the reliable application of BDS-1 satellite time service in the power time synchronization system, which greatly affects large scale application of BDS timing service in the power time synchronization system. BDS-2 adopts 1.5 GHz frequency, which has been put into commercial use in 2013 with less space interference, which is conducive to the large-scale application of BDS timing service in power system.

With the rapid development of GNSS timing service, it has been widely used in electric power, communication, transportation, military and other fields. In the construction of China economy, there are more and more demands for GNSS timing and higher precision in various industries, such as the clock synchronization system of automated power dispatching and communication system in power industry; the time and clock synchronization of switch, access network, transmission network, billing system and network management system in communication network; the clock synchronization of mobile base station, switch and network management system; single frequency wireless coverage in the field of radio and television; command and dispatching system in the field of transportation; unified time system of financial and securities systems, etc. In national defense construction, time and frequency is an important factor related to the victory or defeat of war. The fields of cooperative operations, accurate strike evaluation, military communication, measurement and control, and weapon launching all put forward high requirements for time and frequency accuracy. Weapons, communication systems, command automation systems, and strike evaluation all need high-precision time benchmarks. At present, GNSS time–frequency applications in China have wide demands, but with prominent security risks and weak industry foundation. In recent years, the application of BDS timing service is changing the original situation. Urgent demand for timing service and broad market prospect require BDS application.

The power system is a time-dependent system, and the changes of voltage, current, phase angle and power angle are all based on the time system parameters. The grid connected operation of supercritical units, the interconnection of large regional power grid and UHV transmission technology are all based on a unified time benchmark. The safe and stable operation of power grid puts forward new requirements for power automation equipment, especially for the accuracy and reliability of time synchronization. Relay protection device, automation device, safety and stability control system, energy management system and production information management system are required to operate based on a unified time benchmark, so as to meet the requirements of synchronous sampling, system stability identification, fault location, fault recording, fault analysis and fault inversion. This guarantees the accuracy of line fault ranging, phasor and power angle dynamic monitoring, unit and power grid parameter verification, as well as the level of power grid fault analysis and stability control are ensured to improve the efficiency and reliability of power grid operation.

5.3.2 Key Technologies

The power system needs accurate time synchronization, mainly including the following aspects.

1. Application in fault analysis

Modern microcomputer based intelligent protection devices generally have fault data records or action reports with time scales, which can be used for fault analysis conveniently. If there is no unified time benchmark, the analysis based on these fault records will be meaningless. The adoption of unified time in the fault recorder, event recorder, microcomputer relay protection and safety automatic device, remote control and microcomputer monitoring system installed in the substation (plant) will help to effectively analyze the action and system behavior of various devices during the fault and operation of the power system. The cause and development process of the accident are determined to ensure the safe operation and improve the efficiency of the power system.

2. Application in fault location

At the moment of transmission line fault, voltage transients, i.e. traveling wave, will occur from the fault point to both ends of the line. After the traveling wave signal is monitored and recorded based on the accurate time reference, the accurate fault location can be obtained by analyzing the difference of receiving time between the two sides. Power traveling wave fault ranging and location method is not affected by transition resistance, system parameters, series compensation capacitance, line asymmetry and transformer transformation error, which is an important means of power fault analysis. Because the speed of traveling wave is close to the speed of light, if there is an error of 1 μs on both sides, the distance error is 300 m. The accuracy of synchronous time in power system fault location is better than 0.1 μs.

3. Application in automatic control

Many control strategies in power system adopt timing control strategy, such as automatic reactive power/voltage control. According to the predicted load curve, the dispatcher makes the tap adjustment plan of main transformer and the switching plan of capacitor bank. Accurate time synchronization is particularly important.

4. Application in frequency monitoring

By comparing the difference between the electric clock (also known as power frequency clock) and the standard time, the accumulation of system frequency error is calculated. If the standard time is not accurate, the comparison is meaningless and cannot meet the requirements of power generation, transmission and distribution operation management.

5. Application in phase measurement

Through the voltage and current phase relationship between each node (power station) of the power grid, we can accurately understand the static and dynamic behavior of the power system, carry out reasonable generation and load dispatching, and take targeted stability control measures. The system adopts a unified time reference to synchronize the sampling pulse of input signal of each power station, and accurately measure the phase of voltage and current between power stations. In order to ensure the accuracy of phase measurement, the sampling pulse synchronization error should be as small as possible. If the phase error of power system is better than $1°$, the synchronization accuracy must not exceed 55 μs. At present, the power industry technical specifications require that the time synchronization accuracy is less than 1 μs to meet the requirements of high-precision phase measurement in smart grid.

6. Application in current differential protection

In many kinds of transmission line relay protection, transmission line current differential protection has the advantages of simple principle, high reliability and wide range of application, which is the main technical development direction of line protection. There are two technical difficulties in realizing digital current differential protection: one is the transmission of data at both ends of the line, the other is the synchronous acquisition of data at both ends. The problem of data transmission is easy to solve. In order to solve the problem of synchronous data acquisition at both ends, it is necessary to have high-precision time synchronization.

7. Application in relay protection device test

Line pilot protection (such as high frequency phase difference protection) is installed in the power station at both ends of the circuit. After the system clock is unified, the relay protection test devices at both ends can start according to the predetermined time sequence to generate voltage and current signals to simulate line fault to more comprehensively test the action behavior of the pilot protection device.

8. Application in electric degree collection

It is very important to have a unified high-precision time reference when analyzing the power loss of power grid, which directly affects the accuracy of the analysis results. In the billing subsystem of dispatching automation system, a unified time benchmark is particularly important.

To sum up, the power time synchronization system is an important basis for the safe and reliable operation of the power grid, and is one of the key equipment of the existing power grid and smart grid. The automated power dispatching system, substation computer monitoring system, thermal power plant unit automatic control system, microcomputer relay protection equipment, power fault recording equipment, synchronous phasor measurement equipment all rely on high-precision time synchronization benchmarks. The power time synchronization system provides a unified time benchmark for all levels of dispatching, power plants, substations and centralized control centers to ensure the time consistency and accuracy of sampling information in all aspects of power generation, transmission, transformation, distribution and consumption.

5.3.3 Application Scheme

1. Timing service terminal based on BDS-2

The timing service terminal based on BDS-2 is mainly composed of broadband antenna/preamplifier unit, RF/IF radio frequency unit, digital signal processing unit, frequency synthesis unit, positioning and timing service processing unit, display control unit and power supply unit. Figure 5.8 shows the timing service terminal system framework based on BDS-2.

The timing service terminal based on BDS-2/GPS receives BDS and GPS signals at the same time. The radio frequency signals of BDS-2 and GPS are transmitted to the digital signal processor through the dual-mode receiving antenna of BDS-2/GPS. The radio frequency signals are transmitted to the digital signal processor through the selection filter, low-noise amplifier and image suppression filter. The digital signal processor uses the I/Q sampling signal and clock generated by the RF/IF unit to continuously capture, track, measure pseudo range and demodulate messages from several BDS and GPS satellites, and sends the positioning data and integrity data to the microprocessor for positioning and integrity calculation. The reference frequency of the frequency synthesizer is 5/10 MHz. The required local oscillator and sampling clock are synthesized by phase-locked frequency multiplication and frequency division. The digital signal processor works under the control of navigation positioning and filtering microprocessor, selects the integrity signal of health satellites, selects the satellites according to the best geometric accuracy factor. It provides the best positioning and timing accuracy, obtains the message and measured pseudo range, performs filtering, positioning and timing calculation, and displays the position and time status information. The timing service and timing processing module can continuously and accurately measure the instantaneous time difference between the local time and BDS time, and output high-precision time–frequency reference. The display control module controls the function of the receiver,

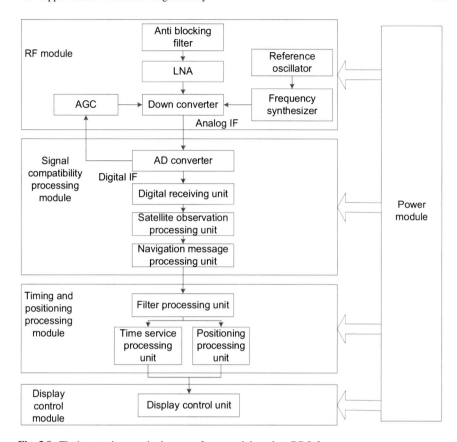

Fig. 5.8 Timing service terminal system framework based on BDS-2

inputs the initial setting parameters, and displays the working status, position, time and other information.

2. The time synchronization system of power network

The three-level time synchronization system architecture of provincial dispatching, local dispatching and substation/power plant is adopted in the time synchronization system of the whole power grid. The E1 link of power SDH network is used to transfer high-precision time reference and automatically eliminate SDH transmission delay. The satellite and ground time reference are backups for each other. High precision ground time reference (accuracy is better than 1 μs) is transmitted between provincial dispatching, local dispatching and substation/station through SDH network E1 link, which provides unified time reference for local dispatching substation and station to realizes time unification of time synchronization system at all levels of power grid.

The time synchronization system of the whole power network adopts the three-level network topology structure of cascade connection, which is composed of the first level time synchronization system (located in the provincial dispatching), the second level time synchronization system (located in the local dispatching) and the third level time synchronization system (located in the substation and power plant). The E1 service channel of the power SDH network is used to transmit the ground time benchmark to realize the time synchronization of the whole grid.

5.4 Application of Emergency Rescue

5.4.1 Introduction

BDS has the ability of two-way message communication between users and ground control center. General users of BDS-2 can transmit 36 Chinese characters at one time. Approved users can transmit 120 Chinese characters at most at one time by using continuous transmission mode. The expanded BDS-3 system can transmit 1000 Chinese characters at a time. This two-way message communication service can effectively meet the requirements of various types of users with small amount of communication information and high real-time requirements. Short message function is very suitable for group users in a large range of monitoring management and data collection in undeveloped areas. BDS will be very useful for users who need positioning information and need to transmit it. It should be specially pointed out that the two-way communication function of BDS is not provided by the widely used GNSS (such as GPS and GLONASS).

Based on the functions of navigation and positioning, short message communication and position report of BDS, it can provide nationwide real-time emergency rescue command and scheduling, emergency communication, rapid information reporting and sharing and other services, which significantly improving the rapid response ability and decision-making ability of emergency rescue.

Combined with the development of emergency rescue technology, BDS can provide services for tourists, outdoor sports, travelers and professional rescue. Using the unique short message position return function of BDS, we can build an outdoor emergency rescue platform based on BDS compatible system, and apply BDS compatible emergency rescue terminal to provide rescue information service without public communication network coverage.

5.4.2 Application Scheme

The outdoor emergency rescue service platform based on BDS includes data center, operation center and customer service call center. The architecture of outdoor emergency rescue service platform based on BDS is shown in Fig. 5.9. The platform is divided into basic layer, data layer, application layer and user layer.

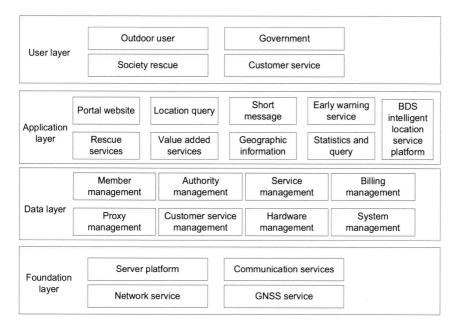

Fig. 5.9 Architecture of outdoor emergency rescue service platform based on BDS

The user can send the signal to BDS satellites by pressing the emergency call button. BDS satellites forward the signal to the ground station. After receiving the forwarding signal, the ground station calculates the location of the user and sends the user information and location information to the data center through the dedicated data line. After the data is saved, the data center will start the basic service module and GIS service according to the distribution rules set by the operation center. The location of the emergency signal is shown in the customer service call center as an alarm with sound and light signal. After receiving the alarm, the customer service call center will provide the location and other information of the user to the rescue agency. The rescue agency will provide rescue services to the user according to the location and other information of the user. Figure 5.10 shows the network diagram of emergency rescue service system.

Each center has the following functions.

1. Integrated operation service center

The integrated operation service center is responsible for the normal operation, configuration management and log management of the main system. The center is connected with the data center, customer service call center and regional centers through the data special line.

2. Data center

The data center adopts the virtual server cluster at the bottom of hardware. The upper layer uses the virtual server modular stacking, including the basic service

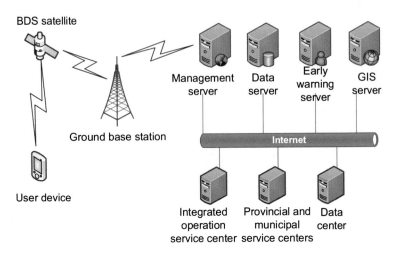

Fig. 5.10 Network diagram of emergency rescue service system

module of emergency rescue and text message, and the value-added service module of footprint recording, updating SNS community status, vehicle management monitoring and ship management monitoring. The data center uses large-scale enterprise database to manage and save user information, map information and related business information, as well as mirror remote hot backup to ensure data security.

3. Customer service call center

The customer service call center uses the special data line to connect with the operation center and the data center. Through automatic voice query, manual agent service and information processing, the customer service call center can provide the confirmation after the emergency rescue and call the rescue organization to serve the users of the emergency rescue terminal.

4. Provincial application service centers

Provincial application service centers provide value-added services to system users. The basic service process of the service platform is as follows:

(1) The user carries the rescue terminal and presses the function key according to different situations.
(2) The rescue terminal automatically sends the distress signal through the BDS terminal.
(3) The BDS receives the distress signal and transmits the distress signal and user location coordinates to the service platform through calculation;
(4) The service platform notifies the rescue information to the emergency contact designated by the rescuers in advance through e-mail, SMS, instant messaging and other channels, and sends the rescue information to the local rescue department at the same time;

(5) The local rescue department shall organize effective rescue according to the returned location coordinates of users.

5.5 Application of Building Safety Monitoring

5.5.1 Introduction

With the wide application of GNSS, more and more high-efficiency, high-precision and portable GNSS devices are applied in the field of civil engineering displacement monitoring. Following the gradual application of BDS in vehicles, ships and other navigation fields, with the introduction of high-precision and GPS compatible BDS receiver, the displacement monitoring of civil engineering is also gradually applying BDS. The design of BDS based survey control network, high-precision displacement calculation and BDS displacement data evaluation method are the key technologies in the safety monitoring industry.

High precision GNSS receiver is installed on large-scale engineering structure to continuously monitor displacement and deformation in real time. It analyzes displacement monitoring and other monitoring indicators (stress, vibration, etc.), masters the operation state of the structure and predicts behavior characteristics, so as to realize the safety early warning and intelligent management of large-scale structure. This displacement monitoring technology based on GNSS has been applied and demonstrated in many fields, such as super large bridges, highway slopes, tunnel construction, mining safety, earthquake monitoring, water conservancy facilities and so on. Through this technology, the high efficiency, high precision and strong adaptability of GNSS measurement are brought into full play, and the real-time measurement of three-dimensional deformation and displacement of large structures is completed. Table 5.1 lists some important projects in which GNSS monitoring has been installed.

It can be seen from Table 5.1 that the application of GNSS in health monitoring of large-scale projects has been very extensive, involving super large bridges, tunnel slopes, water conservancy projects, high-rise buildings and earthquake monitoring. Taking the bridge as an example, because the GNSS measurement does not need intervisibility and can be tested automatically all the time, it is widely used in the bridge operation information management. Especially in the health monitoring system of super large bridge, the displacement monitoring system based on GNSS is studied and installed or prepared to be installed. For example, France tested the Normandy bridge with a total length of 2141 m in 1995, which proved that GPS can monitor the real-time horizontal displacement with centimeter level accuracy. Since 1997, GPS has been used in suspension bridge monitoring in Britain. The vibration displacement of Humber bridge with main span of 1410 m has been measured. The test results are consistent with the model results. In 1998, the experiment of GPS real-time monitoring displacement of Qingma bridge was carried out. When the wind force is level 5, the maximum transverse displacement of the bridge is 64 mm, and the period

Table 5.1 Some important projects of satellite positioning monitoring have been installed at home and abroad

Super large bridge with satellite displacement monitoring

No	Project name	Type	Application
1	Hangzhou Bay Bridge	Cable-stayed bridge	Monitor the deformation of main beam, bridge tower and other components
2	Runyang bridge	Suspension bridge	Monitor the deformation of main beam, bridge tower and other components

Super large bridge with satellite displacement monitoring

No	Project name	Type	Application
3	Huangpu Bridge over the Pearl River in Guangzhou	Cable stayed bridge + suspension bridge	Monitor the deformation of main beam, bridge tower and other components
4	Ningbo fifth road fourth bridge	Cable stayed bridge/arch bridge	Monitor the deformation of main beam, bridge tower and other components

Slope monitoring by satellite

| 1 | Slope of Nanping section of Fuzhou Yinchuan Expressway | Highway slope | Monitoring surface displacement |
| 2 | Huolin River Slope | Mine slope | Monitoring surface displacement |

Construction survey of river crossing tunnel based on satellite monitoring

| 1 | River bottom settlement monitoring of river crossing tunnel | Tunnel | Satellite positioning of survey ship |
| 2 | Zhongtianshan extra long tunnel | Tunnel | Construction survey |

Monitoring dam deformation by satellite

| 1 | Xiaolangdi dam | Water conservancy | Deformation monitoring |
| 2 | Plain reservoir dam | Water conservancy | Deformation monitoring |

Other

1	Hong Kong Polytechnic University building	High rise building	Monitoring wind induced deformation
2	Vertical crustal deformation in Tianjin area	Earthquake monitoring	Monitoring Crustal Deformation
3	Republic Plaza building, Singapore	High level monitoring	Deformation monitoring

is 16 s. The measurement results are in line with the design value. The automatic real-time displacement monitoring system based on GPS of Humen Bridge started operation in 2000. It is the first real-time monitoring system for suspension structure super large bridge in China.

5.5.2 Key Technologies

By using the compatible high-precision displacement monitoring receiver based on BDS/GPS dual-mode, the remote unattended real-time monitoring and early warning network system can be realized. The displacement monitoring engineering of highway infrastructure, long-span bridges and high slopes can be realized using BDS. Figure 5.11 shows the application of GNSS in safety detection.

1. BDS/GPS dual mode compatible high precision displacement monitoring receiver

(1) The BDS/GPS multi frequency high-precision receiver suitable for deformation monitoring of large structures.

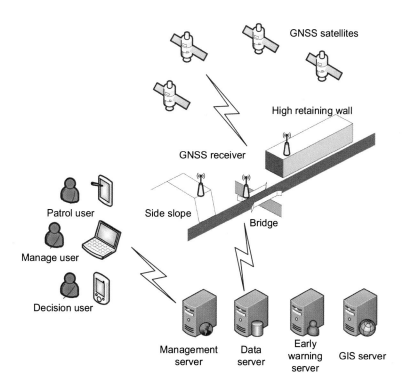

Fig. 5.11 Application of GNSS in safety detection

The OEM module with the ability of receiving and processing BDS/GPS dual frequency signals at the same time should be selected. The corresponding dual frequency measurement antenna should be configured. Considering the data storage and transmission, it should also be equipped with the corresponding large capacity storage unit and mobile communication transmission module. In addition, uninterruptible power supply is needed to ensure the continuity of monitoring.

(2)   Multi path detection technology and elimination method for complex environment of long-span bridge and highway high slope.

In the construction and operation of long-span bridges, due to the large number of steel structure components, multipath effect will become one of the main interference sources in GNSS. Because multipath effect is affected by many factors, such as station environment, constellation, observation time and so on, it is difficult to model accurately. Therefore, multipath effect is generally treated as random error, which is feasible when the accuracy requirement is not high or a large number of observation data (such as one day) is used for data processing and analysis. If the monitoring results reach millimeter level accuracy, appropriate methods must be adopted to obtain precise orbit ephemeris. Double difference method is used to eliminate tropospheric and ionospheric errors. Receiver clock error and high-precision coordinates form joint measurement of international IGS station are obtained to analyze the corresponding multipath error.

(3)   The method of height measurement accuracy of compatible BDS displacement monitoring receiver to millimeter level.

The deformation monitoring of large structures such as high slopes and long-span bridges requires the elevation accuracy to reach millimeter level. Because the elevation accuracy is strongly related to the atmospheric delay error, the elevation accuracy is poorer than the plane position accuracy. Even in static monitoring, it is difficult to reach millimeter level accuracy, and even worse in dynamic monitoring. For the medium and long-distance baseline, BDS/GNSS combined precise single epoch single point positioning algorithm can be used to improve the accuracy and realize the real-time monitoring of millimeter level accuracy in deformation monitoring.

2.   Real time monitoring and early warning system for large structures

The displacement monitoring system of long bridge, tunnel and highway slope needs to take the GNSS as the core and combine with other monitoring indicators to build a real-time monitoring and safety early warning system for large structures. By using the multi index data fusion analysis and evaluation technology dominated by displacement monitoring, the state assessment and safety early warning of large structures can be realized through GNSS. The contents include the following two aspects.

(1)   Displacement monitoring, condition assessment and safety warning of super large bridge structure.

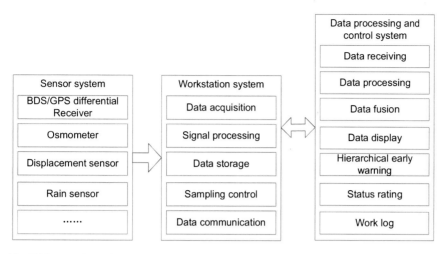

Fig. 5.12 Functions of deformation monitoring system

The functions of deformation monitoring system is shown in Fig. 5.12.

According to the general design code of highway bridges and culverts, bridges with a total length of more than 1000 m or a single span of more than 150 m are super large bridges. There are 689,417 bridges and 2341 super large bridges in China. The establishment of displacement monitoring system for super large bridges includes GNSS monitoring network design, GNSS monitoring system integration, data analysis and safety warning.

The design of GNSS monitoring network mainly includes the selection of key positions and the design of control network. The key position selection is to find the most needed monitoring position of the structure through the finite element calculation. The design of the measuring point control network is to optimize the shape of the GNSS measuring point network to achieve the highest accuracy. Therefore, the content of this section is to combine the two methods of finite element structural analysis and measurement control network optimization. Not only to select the key monitoring points of the structure, but also to make the shape of the monitoring network meet the requirements of the highest monitoring accuracy.

The integration research of GNSS monitoring system refers to the establishment of multi index data fusion analysis and evaluation technology based on displacement analysis by combining structural displacement monitoring with stress monitoring, vibration monitoring and environmental monitoring. Data analysis is to master the displacement change of the structure through the analysis of GNSS displacement monitoring data. Through modal analysis, finite element calculation and other means, combined with stress, vibration and other monitoring indicators, the state of the structure is evaluated. Safety early warning is to give early warning and alarm in time when the evaluation result is unsafe. At present, the transmission of early warning information is mainly through cable, optical fiber and mobile communication network. Under normal conditions, early warning information can be transmitted. However, in the event of major disasters (earthquake and debris flow), when the cable is interrupted, the communication base station collapses and the mobile phone signal cannot be transmitted, the monitoring information is difficult to transmit. By using BDS, important monitoring information can be transmitted through satellite signal, which overcomes the limitations of conventional transmission mode, so as to build a GNSS safety monitoring platform that uses satellite to monitor displacement and transmit information at the same time.

(2)   Monitoring and safety assessment of high-grade highway slope collapse, tunnel surrounding geological disasters and soft subgrade subsidence.

Highway slope collapse, tunnel surrounding collapse and soft subgrade subsidence are frequent geological disasters in China, the main reason is the lack of effective monitoring of soil displacement. The GNSS can effectively monitor the soil sliding, and realize the monitoring of the soil sliding around the highway slope and tunnel. Using BDS compatible high-precision displacement monitoring technology, satellite remote sensing data receiving and processing system and other technologies, we can establish a highway slope, soft subgrade and tunnel surrounding address disaster monitoring system with high reliability and timeliness. Using the dual functions of positioning and communication of BDS, we can solve the problems of slope displacement data monitoring and data transmission.

3.   Infrastructure security operation monitoring center

With the development of information technology, using GNSS, geographic information system (GIS), remote sensing technology (RS) can establish the safety operation monitoring center of super large bridges, important roads, tunnels and other infrastructure. BDS can play the functions of satellite communication, positioning monitoring and navigation monitoring. By installing satellite monitoring devices in bridges, tunnels and other important facilities, they can achieve the displacement monitoring of the structure itself and the remote sensing monitoring of the environment where the facilities are located. Then it can build the geographic information system of important infrastructure through GIS technology and monitor the important highway infrastructure through satellite remote sensing. The monitoring results can be reflect in real time in the monitoring center to achieve the effect of real-time monitoring and security alarm.

5.6 Application of Precision Agriculture

5.6.1 Introduction

Precision agriculture is a combination of agricultural machinery technology and modern spatial information technology under the background of increasing population and decreasing cultivated land year by year, in pursuit of the minimum input, in exchange for the maximum output and the best quality and the minimum harm to the environment. Precision agriculture is a modern agricultural production technology that implements precise timing, positioning and quantitative control of agricultural materials and farming. It can maximize agricultural productivity and is an effective way to achieve sustainable development of agriculture with high quality, high yield, low consumption and environmental protection.

Precision agriculture is developed under the conditions of large-scale operation and mechanized operation in developed countries. In 1995, the United States began to equip GPS on combine harvesters, marking the birth of precision agriculture technology. In the 1990s, precision agriculture was first applied in the United States and Canada, and then in the United Kingdom, Germany, the Netherlands, France, Australia, Brazil, Italy, New Zealand, Russia, Japan, South Korea and other countries. In developed countries, precision agriculture has become an industry combining high-tech and agricultural production, which has been widely recognized as an important way of sustainable development of agriculture. The promotion and utilization of precision agriculture technology has achieved remarkable economic and social benefits.

5.6.2 Key Technologies

China is a large agricultural country. As an emerging technology, precision agriculture has established a certain scale of experimental areas in Beijing, Shaanxi, Heilongjiang, Xinjiang, Inner Mongolia and other places. However, it is still in the stage of experimental demonstration and breeding development, especially in the high-precision agricultural machinery precision control system products, which has been relying on imported products for a long time, seriously restricting the development of precision agriculture development in China.

1. Precision control of agricultural machinery

Precision control technology of agricultural machinery mainly includes the following three aspects.

(1) Accurately control the driving route of agricultural machinery. The automatic driving function of agricultural machinery is realized through the precise position provided by GNSS. Using GNSS and automatic driving technology of agricultural machinery, the driving path of agricultural machinery can be accurately controlled. The real-time positioning error can reach centimeter level, which ensures the repetitive operations of ridging, sowing, fertilizing, spraying, irrigation and harvesting. The unmanned operation can work normally even at night. It greatly improves the efficiency and effect of farmland operation. Due to the adoption of automatic driving technology, the driving and operation requirements of agricultural machinery are reduced, and the driver workload is reduced.

(2)   Accurate control of farm tools operation. Through the hole control and flow control of sowing, fertilization, sprinkler irrigation and other farm tools, the variable rate application of farm tools is realized. In the process of agricultural operation, sensors are used to monitor the soil moisture, crop growth, distribution of diseases and pests, historical yield and other information of farmland for a long time. Combined with BDS spatiotemporal data for analysis and calculation, the grid farmland state distribution map is generated. Each farmland is targeted for fine operation management to explore the maximum potential of farmland, which can be adjusted according to the characteristics of farmland. The adjusted input of pesticides and fertilizers has reduced the environmental pollution to a certain extent.

(3)   Integrated management scheme for manufacturers. GNSS based automatic driving products of agricultural machinery are most widely used in precision agriculture. Many manufacturers around the world provide automatic driving products of agricultural machinery, including Novariant, Trimble, Hemisphere, Topcon, Leica, etc. The composition of agricultural machinery automatic driving products can be divided into two parts: GNSS receiver and automatic driving control device. According to the different characteristics of crops and farms, there are many ways of integrated management.

The carrier phase differential positioning technology is used in the GNSS receiver, which effectively improves the positioning accuracy. The real-time positioning accuracy can reach centimeter level, ensuring the repeated accuracy of farmland operation. According to different operation accuracy requirements, users can choose the appropriate GNSS receiver.

The automatic driving control device can ensure that the agricultural machinery operates according to the planned path. According to the requirements of the operation, the user can require the agricultural machinery to drive along a straight line, a circle, a specific curve, an intelligent obstacle avoidance, and an automatic path planning within the effective plot. According to the current position and posture of agricultural machinery, the automatic driving control device analyzes and calculates the trajectory error by combining with the navigation planning path, and corrects the heading of agricultural machinery in time by controlling the driving and steering system of agricultural machinery. The control mode of automatic driving control device for agricultural machinery can be divided into mechanical control, hydraulic control and can bus control. Different control modes should be different according to different types of agricultural machinery and different applications.

2.   Variable rate application control

The basis of variable rate application is the information collection and data analysis and processing of farmland, relying on automatic measurement tools such as temperature and humidity detection, soil composition detection, spectral analysis, and even manual measurement, combined with GNSS data, to establish a detailed farmland state map, and carry out variable operation on this basis.

Manufacturers engaged in the research of variable operation control technology include Novariant, Trimble, Leica, Topcon, AGLeader, Raven, TEEJET, Dickey-John, etc. They provide rate application control products including a variety of data acquisition sensors, data analysis software, variable seeding and fertilizing machine, variable rate spraying machine, etc. This variable rate application control product can directly connect with the on-board computer of the automatic driving system of agricultural machinery. During the process of automatic driving, it collects farmland information or carries out variable work according to farmland state map. It sows seeds according to the shape of the plots, determines the amount of chemical fertilizer according to the soil moisture content, and selectively sprays the pesticide according to the degree of crop diseases and insect pests, thereby saving the input of agricultural materials. It also reduces the pollution to the environment to a certain extent.

3.   Integrated management program for agricultural organizations

In the United States and other developed countries, precision agriculture has developed into a comprehensive solution combining spatial information technology with agricultural production. Trimble is a leader in precision agriculture, GPS and navigation solutions in North America. It can help users operate vehicles and machines more efficiently, save money, and increase production and productivity. Connected Farm™ of Trimble is a comprehensive management scheme that can provide real-time information transmission for the whole farm. Connected Farm™ can manage your farm in real time by simply accessing it through a browser. You can collect information through your smartphone (download relevant apps from Trimble website) or Trimble vehicle terminal, and upload it to your associated farm account. Login farm account through the browser, you can easily view, sort and print relevant information, including vehicle location data, engine performance data, vehicle alarm situation, field boundary information, etc. Users can make management decisions in any place, and the system will back up the data to ensure the safety and reliability of the data. However, Connected Farm™ management scheme is more suitable for the family farm mode, and cannot solve the problem of large-scale cross regional operation and local operation scheduling of agricultural machinery.

5.6.3 Application Scheme

In order to improve the accuracy of ridging and sowing, and improve the planting environment and conditions of crops, BDS CORS station was established in the farm, and BDS ground based augmentation system network was set up to provide centimeter level RTK differential correction service for precision planters. Deploying GNSS terminals for monitoring, scheduling and navigation on agricultural machinery can improve the efficiency of agricultural machinery monitoring and scheduling. Using data acquisition GNSS terminal and pad in farm for wireless collection, transmission and management of agricultural operation information, as well as enhanced mobile management of farmland can improve the efficiency of farmland information collection and the level of informatization and intelligence, and improve the management efficiency and automation level of agricultural production organization.

1.   Precision control of agricultural machinery

Precision control of agricultural machinery mainly includes the following three parts.

1)   Development of high precision measurement equipment based on BDS

Based on BDS RTK Positioning Technology, combined with GPS and GLONASS integrated navigation and positioning technology, the multi-mode multi frequency RTK high-precision measuring equipment, including RTK reference station and mobile station, is developed for real-time position measurement of agricultural machinery. This equipment can provide space scale for agricultural machinery operation, make farmland ridging, sowing, fertilization, spraying, irrigation and other operations more accurate. It ensures the repeated precision of farmland operation, maximizes the land utilization rate, reduces the waste of agricultural materials, reduces environmental pollution, and realizes the fine management of farmland.

2)   Precise control management equipment for agricultural machinery

Precise control management equipment for agricultural machinery is the control, calculation and management center of agricultural machinery precision control system, which is responsible for the task management, navigation path planning, configuration of positioning measurement equipment and working parameters of mechanical control equipment of the whole system. It coordinates the collaborative work of all parts of the system. At the same time, Precise control management equipment for agricultural machinery is also the human–machine Interaction of agricultural machinery precision control system. It provides a visual operating environment for users. The operator can master the current system task and working state at any time through the equipment, and can also input control commands to the system through it.

3)   Precision automatic control equipment for agricultural machinery

Precision automatic control equipment for agricultural machinery, mainly agricultural machinery automatic control device, applies the gyroscope and angle sensor monitoring of agricultural machinery attitude and movement in real-time. The automatic adjustment of agricultural machinery hydraulic steering system ensures that the path of agricultural machinery in line with the navigation path planning. The equipment can not only ensure the accuracy of agricultural machinery movement, achieve high-precision repetitive operation, but also reduce the work intensity of the driver. The operator has more energy to pay attention to the growth of crops and other information to make night operation possible, which greatly improves the efficiency of farmland operation.

2.   Management scheduling for agricultural production organization

1) Multi level agricultural machinery operation management and scheduling

Stable agricultural machinery operation service mode is the foundation of the successful application of spatial information technology, and also the advantage of the application of spatial information technology. The large-scale cross regional operation service mode of agricultural machinery has gone through three main stages: C2C (business mode between individuals), C2B (business mode between individuals and collectives) and B2B (business mode between collectives), and will gradually enter a new stage of combining cloud organization and self-organization.

Chinese agricultural machinery operation management system can be divided into three levels, namely the competent department level, agricultural production organization level and agricultural machinery operator level. From bottom to top, agricultural production organizations monitor, manage and dispatch agricultural machinery operators and agricultural machinery under their jurisdiction. Accordingly, the competent department can supervise and serve the agricultural production organizations under its jurisdiction.

2) Agricultural machinery operation management and scheduling system

Based on spatial information technology and wireless communication technology, the agricultural machinery operation management and scheduling system uses various mobile intelligent terminals to establish data interaction networks between different levels, offices and fields. The system obtains farmland information, makes scientific decisions, and dispatches agricultural machinery for field operation. The structure of agricultural machinery dispatching system is shown in Fig. 5.13.

Agricultural machinery operation management and scheduling system for agricultural production organization mainly includes the following specific contents.

(1) Agricultural production information wireless collection and management subsystem. The system uses GNSS for real-time positioning, collecting spatial information and corresponding attribute information (such as farmland electronic map, main road, gas station, grain depot, etc.), collecting image of corresponding location, collecting farmland, crop and other attribute information

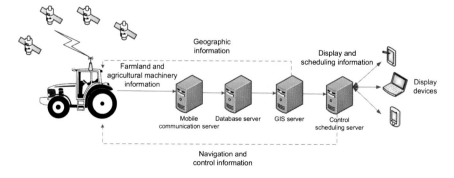

Fig. 5.13 Structure of agricultural machinery dispatching system

and production information, realizing data management and storage based on local database, and real-time synchronization with remote data between central database.

(2) Agricultural machinery operation intelligent decision-making and spatiotemporal scheduling subsystem. The problem of agricultural machinery and farmland planning and distribution can take many forms in reality. In terms of quantity proportion, the proportion of farmland to agricultural machinery can be one to many, many to one, many to many. In terms of operation content, it can be sowing, fertilizing, spraying, harvesting, etc. In essence, this kind of problem can be classified as agricultural machinery spatiotemporal scheduling problem, that is, to complete the given task with the highest efficiency (the least cost) in the specified time and region. Based on GIS, spatial database and other technologies, the agricultural machinery spatiotemporal scheduling system can realize data entry, model operation, Gantt chart generation, scheduling instruction generation and other functions.

(3) Agricultural machinery state automatic recognition and accurate statistics subsystem. The whole process of agricultural machinery operation can be divided into four main working states: parking, transfer, operation and transition. Using the information of agricultural machinery longitude, latitude, elevation, speed, direction, time and the number of available satellites reported by the agricultural machinery vehicle positioning terminal, it can monitor and judge the various states of agricultural machinery in real time, realize the automatic extraction of computer information, realize the automatic monitoring of agricultural machinery management, operation statistics and road network update.

(4) Agricultural machinery mobile monitoring and central navigation subsystem. The vehicle terminal regularly reports the position to the server. The mobile monitoring terminal uses a timer to obtain the latest position of agricultural machinery at a certain time interval, locally refreshes the map area, displays the position of agricultural machinery on the map in real time, and infers its operation status according to the speed of agricultural machinery, and displays it in different color icons. Managers can understand the position and status of agricultural machinery in real time and intuitively. According to the need of operation task, the mobile commander selects the destination of agricultural machinery operation on the map and sends it to the vehicle navigation communication terminal. After receiving the center navigation command, the vehicle terminal extracts the longitude and latitude of the destination, generates and displays the optimal path on the terminal, and then guides the agricultural manipulator to the destination by voice navigation.

Chapter 6
"BDS + IOT" and Industry Reform

6.1 Application of Municipal Pipe Network Industry

6.1.1 Introduction

In the field of smart city management, municipal pipe networks such as water, electricity, gas, heat and other basic services are the urban lifelines related to people's livelihood. They are urgently need the accurate location information. With the expansion of urban scope, the length of urban pipeline continues to extend. The coverage area is becoming larger and larger. Many factors such as complex and changeable urban environment cause a large number of potential safety hazards, which are widely distributed and difficult to find and deal with. This brings great challenges to the operation and management of enterprises related to urban lifeline, and often requires a lot of human and material resources to maintain operation. With the continuous improvement of BDS, BDS precise location service makes more and more urban lifeline management enterprises develop from using traditional management methods to using more intelligent comprehensive management. By combining BDS precise location service with daily business applications, many difficulties in operation management can be improved.

In recent years, various types of municipal pipe network management accidents occur frequently, many major safety accidents have caused extremely bad effects, seriously damaged the property and life safety of the country and the people. It is urgent to carry out more refined management from the planning and construction, operation and maintenance and other aspects. The application of BDS precision service is imminent. The "BDS precision service network" promoted by GLAC has covered more than 600 cities in China, providing application services for gas, drainage, heating, electric power and other industries. At present, based on the construction and operation of the "BDS precision service network", especially for the municipal pipe network industry, Beijing CNTEN Co., Ltd. has provided the whole life cycle BDS precision service of pipe network from the beginning of construction, focusing on the application of urban gas industry.

© Publishing House of Electronics Industry 2022
B. Wang et al., *Internet of Things and BDS Application*,
https://doi.org/10.1007/978-981-16-9194-2_6

6.1.2 Application Scheme

With the rapid economy development of China, the underground pipe network industry, which is closely related to the safety of people's life and property in the city, has been paid more and more attention by the whole society. Among them, the gas pipeline network has received special attention due to its special properties in the field of safety. At the same time, the original underground pipeline network data of chained and independent gas company in China is difficult to match the application demand of higher accuracy.

1. Gas pipe network construction management

The application of BDS precision service network in the field of gas is precisely through the simple and easy-to-use intelligent precision positioning terminal. Each link of gas application can obtain accurate spatiotemporal information to continuously improve and correct the underground pipe network data. In each link, as the source and initialization stage of pipe network data acquisition, the collection and management of accurate construction data is particularly important.

By introducing the BDS precise location service into the pipeline network construction process, centimeter level precise location information can be collected and applied at any time in the construction survey, engineering lofting, embedding location, welded joint positioning, attribute return and pipeline network retest. Through the development of intelligent terminal and customized app, the backstage GIS and various application systems can obtain the first-hand information of on-site construction in real time and accurately to ensure the efficient development of pipe network construction and the effective collection of accurate data. Figure 6.1 shows the service platform for pipe network construction.

2. Searching of Gas pipeline components

For a long time, in the process of underground pipeline reconstruction and emergency repair, how to quickly and accurately determine the location of key components such

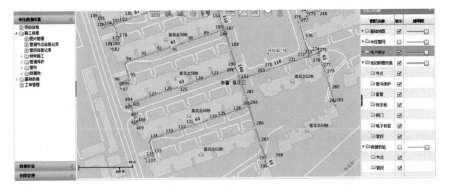

Fig. 6.1 Pipe network construction service platform

as underground pipelines, pipe fittings and welded junctions has become a major problem in the field operation. On the one hand, it is due to the lack of accuracy and reliability of the original underground pipeline network data. On the other hand, it is directly related to the lack of on-site positioning and parts searching means.

Under the premise of ensuring the accuracy of GIS information in the pipe network, the centimeter level precision search service provided by the BDS precision service network compares the personnel position with the component position in real time and provides navigation. The service gives a sound prompt within one meter above the component, which greatly shortens the search time of gas companies in the field operation, saves a lot of personnel and engineering costs, and improves the current efficiency. The efficiency of field operation and the effectiveness of management decision-making is highly improved.

3.   Gas pipeline inspection

Gas pipeline inspection refers to the regular inspection and inspection of the gas pipeline within its jurisdiction by the gas pipeline management department, so as to ensure the safety of gas transmission and prevent gas stealing and leakage. At present, many gas companies have used various information management systems. But due to the particularity of gas pipeline inspection, it needs field operation. The information management of gas pipeline inspection is almost a blank. Generally speaking, gas pipeline inspection has the characteristics of large working area, long line and complex environment, which puts forward higher and higher requirements for the supervision of gas pipeline inspection.

For a long time, the gas pipeline inspection is carried out by manual record or electronic information tag (button), which cannot meet the utility guarantee requirements of real-time supervision and summary analysis. After the BDS precision service network is applied in the gas industry, BDS precision location service is used to combine all kinds of pipeline inspection business processes. The inspection monitoring personnel can compared with the pipeline location background in real time, and accurately analysis line inspection arrival rate. On-site events can be collected and returned back in real time to achieve an efficient, accurate and quantifiable inspection mode.

4.   Gas leakage detection

The city gas pipeline is distributed in the underground of the city. The leakage will cause huge economic losses and personal injury. Timely detection and accurate location of the leakage point has become the primary task of gas leakage detection. The traditional leakage detection mainly uses hand-held detection, road portable inspection and vehicle laser detection to determine the combustible gas concentration information nearby to determine the potential leakage situation. The data attribute is relatively single, which cannot carry out quantitative model analysis after long-term data accumulation, and cannot accurately reflect the overall leakage situation of the pipe network.

Since the application of BDS precision service network to gas management, the leakage detection and monitoring equipment has been reformed. The BDS precision

Fig. 6.2 Precise gas leakage
detection

location service has been integrated into the pipeline leakage detection business. The automatic matching of detection data with sub meter and centimeter precision location has been realized, which greatly reduces the detection blind area, increases the detection probability of daily minor hidden dangers, and avoids secondary accidents. At the same time, BDS precise location service "activates" the historical data of gas leakage detection. Through data fusion and intelligent analysis and calculation, it can achieve the overall safety status assessment of safety monitoring in time and space, and provide intelligent technical support for timely detection of hidden dangers in the pipeline network. Figure 6.2 shows the accurate gas leakage detection.

5.   Gas anticorrosive coating detection

Buried pipeline is an important part of the pipeline. Due to the complexity of buried laying and geographical environment, it is not suitable to use conventional methods for inspection. With the passage of time, under the influence of construction, soil corrosion, land subsidence and other factors, the anticorrosive coating of the pipeline will be aging, brittle, peeling off, causing corrosion perforation of the pipeline, resulting in leakage. Pipeline anticorrosive coating management is the key issue. It is an important part of the operation management of a gas company.

The BDS precise location service provided by the BDS precise service network is integrated with the original pipeline detector, ground penetrating radar and other anticorrosion detection means. On the one hand, it can enable the field operators to quickly determine the accurate location of pipeline valves, cathode piles and other infrastructure facilities according to the accurate location information of GIS. On the other hand, it can record and inspect the accurate location of the anticorrosion layer at the same time. Through automatic matching on site or in the background, every anticorrosion abnormal point is accompanied by accurate location information, which provides solid and accurate data support for excavation maintenance and troubleshooting.

6. Rapid deployment of gas emergency rescue

The main component of natural gas transported by urban gas pipeline is methane. The minimum and maximum concentration of combustible mixture that can explode are called lower explosion limit and upper explosion limit respectively. When the content of methane in the mixed gas exceeds the threshold, it is necessary to organize personnel evacuation.

The rapid deployment system of emergency rescue is specially used for the rapid disposal of hazardous chemical emergency accidents. It provides a variety of monitoring tools for the decision-makers and rescuers of emergency accidents, provides a full range of concentration distribution of dangerous gases and chemicals, and releases control information, meteorological information and vital signs data of rescuers.

The rapid deployment system of gas emergency rescue, combined with BDS precise location service, rapidly deploys the corresponding explosion-proof system at the gas leakage emergency site. The system provides the first-hand data information of precise location for the command center and field operators, and realizes the real-time monitoring of leakage concentration at each monitoring point on the emergency site. It establishes a complete set of regional monitoring system and ensure the personnel safety and operation safety of the emergency site. When the leakage concentration of the monitoring point exceeds the limit, the composite gas detector can timely and effectively give an audible and visual alarm, which can effectively guarantee the safety of emergency site personnel and operation safety. Figure 6.3 shows the rapid deployment site of gas emergency rescue.

Fig. 6.3 Gas emergency rescue rapid deployment site

6.2 Application of Elderly Care Industry

6.2.1 Introduction

The aging level in China is already very high. The phenomenon of "getting old before getting rich" means that the resources that the society can provide are very limited, and the preparation of all aspects is not enough. The rapid development of population aging is having a great and profound impact on the economy, society, even individuals and families.

There are two kinds of traditional pension models in China: family pension and institutional pension. The proportion of Chinese elderly living in pension institutions is about 1%. Due to the family concept formed by traditional culture, the vast majority of the elderly have a special preference for family support in their choice of pension mode. Providing for the aged at home is not only convenient for their children to fulfill their supporting obligations, but also beneficial for the elderly to enjoy family happiness. But the trend of family miniaturization leads to the deficiency of family pension function, there must be social services to provide support.

China is building a social service system for the aged, which is based on home-based care, community-based care, supported by pension institutions, matching financial guarantee with service guarantee, and combining basic services with selective services. It will basically realize that everyone can enjoy pension services. In actively promoting home-based care services, social groups and enterprises are encouraged to engage in home-based care services. Relying on LBS, Internet and IOT, we set up a platform for providing elderly care services by using modern communication and information, computer network and other intelligent control technologies. This is like a "nursing home without walls" to provide safe, convenient, healthy and comfortable services for the elderly.

The national plan for the construction of social pension service system (2011–2015) clearly states that "strengthening the construction of social pension service system is an effective way to expand consumption and promote employment". The huge demand of the elderly for care and nursing is conducive to the formation of the elderly service consumption market. It is estimated that the potential market scale of nursing services and life care for the elderly in China is more than 450 billion RMB, and the potential employment demand of the elderly service industry is more than 5 million.

6.2.2 Key Technologies

Based on the BDS, wireless communication technology, IOT, RFID technology, using BDS compatible module, combined with the elderly needs for family services, emergency assistance, health detection, home care services and other aspects, we can integrate the elderly care operators, elderly care service providers, rescue agencies, medical institutions, etc. And we create a set of modern care services for the elderly. The overall service solution consists of two parts: intelligent location service platform and BDS compatible personal terminal.

The modern elderly intelligent location service platform with spatial location cloud service and spatial decision system provides all the modules needed for the service, including geographic information service, terminal management, user management, community service management, call center, medical expert system, etc. Local elderly care service institutions integrate local elderly care and community service resources. They provide visual image operation interface, and use this platform to provide emergency rescue services, health parameter detection and statistical analysis, home-based elderly care services and community services for the elderly. Through this platform, the family members of the elderly can know the location information and physical condition of the elderly at any time, and provide family care services such as protection circle setting and voice reminder. The user terminal provides one button call, one button emergency help, voice reminder and wireless collection of health parameters for the elderly. At present, the "Guanhutong" elderly care service platform and BDS based terminal promoted by China Unicom, are covering more than 20 provinces and autonomous regions in China, providing elderly care, health and social services for hundreds of thousands of elderly people.

1. Intelligent location service platform

The core system of intelligent location service platform is composed of application service, terminal access service and data storage system. At the same time, it is connected with map service, content service and other systems through content and service standard access interface layer. The overall architecture of the system is mainly divided into three layers: access layer, business layer and interface layer. Figure 6.4 shows the system framework of the service platform.

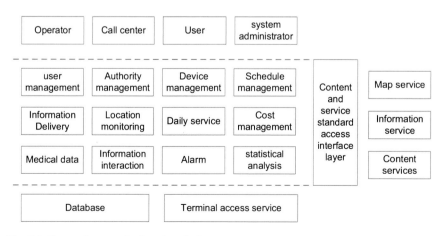

Fig. 6.4 System framework of service platform

Through the three-layer system architecture design, it can effectively support the changes introduced by the increase of business functions, and effectively isolate the changes of a certain level. It reduces the changes of a certain level which lead to the changes of other levels, and reduces the coupling between different levels. Through the support system expansion, such as by supporting multiple versions of the terminal protocol, it can support different terminals.

The application service of the service platform is composed of information interaction, information release, location monitoring, schedule management, expense management, statistical analysis, user management, authority management, equipment management and other business modules. According to the business needs, it provides services to the operation manager, call center service agent, individual user, system administrator and other roles of the operation service provider through B/S mode.

2.   BDS compatible personal terminal

BDS compatible personal terminal adopts modular hardware design, uses MCU to schedule tasks, and adopts positioning module supporting BDS/GPS/GLONASS to ensure that available satellites of different systems can participate in joint positioning. The built-in 3D motion sensor can detect the amount of movement of the terminal holder, and with fall detection, posture detection, etc. When the elderly fall, it can send out alarm information in time. The wireless transmission module is embedded in the base, which is connected with the terminal through the serial port to form a wireless medical gateway. It is responsible for collecting the data of wireless medical equipment and transmitting it to the service platform through wireless communication. The service platform can connect with community hospitals and other medical institutions to feed back the health status of the elderly to doctors in time. If there is an emergency, it can be handled in a very timely manner. Figure 6.5 shows the structure of personal terminal.

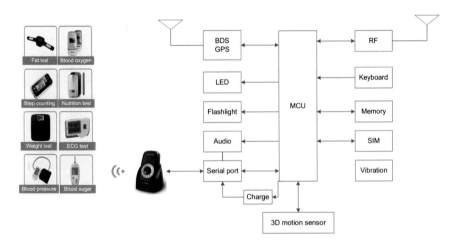

Fig. 6.5 Structure of personal terminal

6.3 Application of Child Care Industry

6.3.1 Introduction

Cognitive ability of children is poor, and they are easy to get lost. In the case of child abduction and trafficking, guardians cannot prevent it. In case of danger, there is no way to inform parents in time. Children are playful and mischievous, and guardians do not have time to pay attention to children all the time. These are the main reasons for children safety problems. Expanding the scope of child protection is in line with the development trend that the international community has basically unified understanding of children issues. According to statistics, the number of abnormal deaths of primary and secondary school students in China each year reaches 16,000, which means that on average one class of students disappears every day. Safety education is a heavy topic. The state and all levels of society attach great importance to it and adopt different ways of safety management and safety education. The last week of March every year is designated as the national "safety education day". When children lose safety, parents lose everything. The sample survey shows that 95% of the parents care order about their children: the first is safety, the second is health, and the third is achievement.

The definition of children safety should be that children should be protected from infringement in all aspects of social life, such as the body, spirit and Internet, and get special protection. The current situation of children safety in China is not optimistic, which is mainly manifested in the high mortality of children caused by accidental injuries, the frequent occurrence of injuries in school life, the existence of domestic violence and so on. The main reasons for safety problems of children are the incomplete understanding of safety concept in traditional society, the misunderstanding of only child education, the neglect of life and rights education, and the lack of preparation for the Internet and the resulting safety problems. Children safety is a systematic project. While establishing a new concept of children safety, further strengthening life education, strengthening the construction of school management system, establishing a new concept of family education, and learning from the experience of developed countries, we should also strengthen the protection of children safety through the Internet, IOT and other innovative scientific and technological means.

The collection of location information of children is not only the basis of children safety protection, but also the key to realize children care. In addition to military use, BDS also plays an important role in economic construction and accelerating the intelligent process of human work and life. BDS integrates navigation and communication capabilities, and has the functions of "real-time navigation, fast positioning, accurate timing service, position report and short message communication service". The service scope and activity track of children are closely related to the precise time and location. BDS plays an important role in accelerating the initiative and pertinence of promoting the cause of children safety.

6.3.2   *Application Scheme*

Based on the BDS, wireless communication technology, IOT technology, RFID technology, combined with parents' demand for children accurate location, emergency rescue, safety warning and service command and scheduling, the kindergarten, primary school, rescue institutions, medical institutions and so on are integrated to provide services to children and parents in the form of service network and call center. The comprehensive safety protection service can realize the integration of children, parents and local children safety operation agencies. The goal of the child care application system based on BDS is to build a system that can meet the needs of children monitoring data collection, transmission, storage, analysis and processing with the principles of reliability, security, scalability, manageability, advanced nature and practicality. The overall architecture of children care service platform is shown in Fig. 6.6.

As a satellite positioning system with completely independent intellectual property rights and its unique technical advantages, BDS has irreplaceable advantages in information security and operation reliability, which can effectively guarantee the security in the process of personal information collection, processing and transmission. Therefore, it is an inevitable choice to collect positions from BDS to achieve positioning requirements. The main functions of children care service platform are divided into five parts: resource access, cloud processing center, operation center, access interface and device access.

The basic positioning data of resource access includes unified user data and warehouse application data. The positioning terminal carried by children receives BDS signal and sends relevant location and time information to the platform, and the platform continuously tracks children activity. The third party resource access, including map providers, content providers and service providers.

The cloud processing center adopts the "Cloud + End" system construction mode to build an integrated children care application platform. "Cloud" is a unified private cloud platform, including infrastructure as a services (IaaS), platform as a services

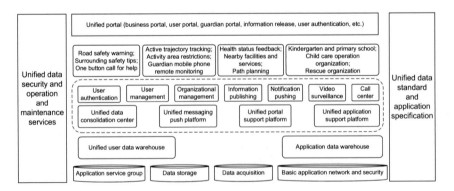

Fig. 6.6 Overall framework of children care service platform

(PaaS) and software as a services (SaaS), which can fully meet the needs of system positioning infrastructure services and software application services. "End" is the application end. Every child and parent is the application end of the private cloud platform. They store and process the collected data and event information, and integrate and optimize the third-party resources to provide services by using information technology, monitoring, analysis and intelligent response.

The operation management, service management, personnel management, call center and billing management of operation service in operation center can realize visual monitoring of workflow progress, information query and monitoring, historical data analysis, and collaborative analysis of relevant experts.

Access integrates the IOT and communication technology into all aspects of children guardianship, integrating life, entertainment and security. It provides children guardians, kindergartens, primary schools, children care operators and government departments with access to information through a variety of platforms (computers, mobile phones, etc.).

Device access is responsible for the access and authentication of personal location terminal, collecting terminal location information, alarm information data, and sending instructions between platform and terminal. It can select whether to access the device according to the user terminal, and the system will match the corresponding service according to the user terminal selection.

6.3.3 Application Function

Integrating the specific needs of children life, using the positioning function of BDS, we can provide services in the areas of children location, safety warning, nearby facilities, one click help, service command and scheduling, etc.

1. Children location service

The positioning terminal device carried by children can provide high-precision continuous spatial coordinate data for guardians in an all-round way through the accurate service information provided by the BDS precision service network. The device can realize the function of all-weather tracking and monitoring children, and regularly transmit the obtained location information to the location service platform through the wireless communication channel. The guardian of the child can access the intelligent location service platform through the computer or using the smartphone App to obtain the location information of the child at any time, and carry out location monitoring, protection circle alarm and other family care services. When the children position does not change or changes abnormally for a long time, the application platform will automatically send an alarm and location information and dial the phone to inform the family and relevant service personnel. When children leave the set protection area, their families will receive alarm and location message. This function is particularly important to prevent children from missing, and can help parents find the lost children in time to avoid tragedy.

2.   Safety alert service

The advantages of BDS based child care application are real-time positioning, knowing children position at all times, setting children activity area and access fence, etc. The terminal can record the children movement track all day long to understand the children activity area. When there are vehicles around the children, it will give a prompt according to the traffic rules when crossing the road.

3.   Nearby facilities services

According to the children location, the device can determine the nearest service providers in the area, such as schools, libraries, playgrounds, hotels, rescue agencies and medical institutions, and plan the path for them. Intelligent navigation can be carried out according to the requirements to guide children to their destination more clearly.

4.   One click help Service

In the past, when children were in danger, the police could not receive the information until their parents or enthusiastic people called the police. They also needed to investigate and inquire about the time and address of the crime. It is often difficult for people to identify the specific location, which will delay valuable rescue time. When a child is in danger, the service terminal of BDS can trigger SOS button to send alarm information to the guardians' mobile phone quickly, and monitor the surrounding sound environment of the child remotely to find the lost child in time. When children need help, they can press the service button on the terminal to dial the service phone of the operation organization and call the agent of the call center. The operation organization will arrange corresponding service personnel or volunteers to give help.

5.   Command and dispatch service

When children need any service, they can press the service button on the terminal to call the operation organization. When the call center agent receives a call from a child, it will get the current position of the child on the intelligent location service platform. The agent can search the service provider from near to far on the location service platform according to the needs of the child. After the children or guardians choose, they will receive the orders from the designated service providers, and the service providers will confirm the orders and arrange the service personnel to provide the corresponding on-site services. After completing the service, the service personnel can inform the children care operation organization according to the specific BDS and wireless communication receiving terminal. Through the intelligent service platform, the agents of children care operators can visually monitor the whole process of service, and can see when the service personnel start, arrive and complete the task. If the service provider or staff fails to respond to the task in time, the agent can reselect the service provider and staff to ensure that children can enjoy high-quality service in time.

6.4 Application of Urban Management

6.4.1 Introduction

Since 2004, Dongcheng District of Beijing has applied modern information technology, combined with 3S (GNSS, GIS, RS) technology, optimized the management process, explored "grid city management mode". A digital city management platform is built, which significantly improved the efficiency of urban management and government management level. The frequent problems in urban operation are well solved, and remarkable results are achieved. The new mode adopts "ten thousand meter unit grid management method" and "urban component event management method", which realizes the refinement of management area and management responsibility, and creates a management system of supervision separation, so that all kinds of urban management problems can be effectively solved. In the aspect of information collection, the urban management system is innovatively developed, which changes the traditional manual recording and telephone reporting mode, realizes the real-time transmission of information, and improves the efficiency of problem discovery.

Grid urban management mode has brought great changes in the field of urban management, and virtually promoted the development of the whole urban management industry. Its role has been fully reflected in Dongcheng District of Beijing, Hangzhou, Changzhou, Chengdu, Shijiazhuang and other cities in China. At present, in the practice of more than 200 cities, the grid management mode has been proved to play a great role in other fields. Jianguomen sub district of Dongcheng District applies the concept of grid management mode, combines with the specific requirements of social management innovation. A number of geographic information data, such as basic terrain data, satellite remote sensing data, urban 3D data, urban real image data, etc., are integrated to build the first grid social management service platform in China. This realizes the transformation of grid management mode from urban management to social management, which expand the field of management services. Ordos city and Changji city actively explore the grid management mode for the management of underground pipelines and facilities, so as to realize the ground and underground integrated supervision. Ningbo city introduces advanced technologies such as video intelligent analysis, management resource positioning and intelligent scheduling into digital city management, realizing the transformation from "digital city management" to "smart city integrated management". Grid urban management has been popularized all over the country, which promotes the development of the whole urban integrated management industry and leads a new direction of the industry.

Urban management departments have been trying to improve the level of urban management and service through various information technology means. However, due to the limitations of the system such as the division of functions, the construction of urban management informatization often falls into the situation of spontaneous and fragmented. Its implementation effect is relatively limited. In order to speed up the transformation of urban industry, enhance the comprehensive competitiveness of the city, and realize the sustainable development of the city, the urban management informatization construction is often in a situation of spontaneous and fragmented. It is difficult to meet the needs of operation and service brought by the rapid expansion of the city.

The continuous deepening of urbanization has brought rare opportunities to urban development, but also brought severe challenges to urban planning and management, social stability and security, and urban sustainable development. Livelihood problems need to be solved, and urban operation and service level need to be improved. In the case of increasingly prominent problems in urban management, we should not only actively promote the construction of digital urban management information system, but also make use of new technologies, new ideas and new ideas to deepen the research of urban management platform. The new information technologies such as satellite remote sensing, GNSS, IOT and cloud computing should be integrated into the construction of smart city. A set of information acquisition and sharing, information processing, whole process monitoring and supervision, analysis and decision-making, video monitoring, emergency linkage, joint command and dispatching, etc. are integrated into an intelligent, full coverage and whole process comprehensive urban management platform. Based on this platform, various resources in the city are highly shared, and various business units are coordinated. Rapid response and accurate management, as well as overall command and whole process supervision and assessment are achieved.

6.4.2 Application Scheme

Based on the establishment of a new mode of intelligent urban management, smart urban management adopts a variety of emerging technical means to achieve the refinement and intellectualization of urban management objects and subjects. In terms of urban management objects, digital and intelligent identification and management are carried out by using component event partition method and IOT technology. In terms of urban management subjects, digital and intelligent identification and management of personnel and various management resources are realized by using handheld, vehicle and other terminal devices. In terms of management process, a series of intelligent means such as IOT and video intelligent analysis are used to realize digital and intelligent identification and management. This process can improve management efficiency and reduce personnel consumption. To sum up, according to the needs of urban management and supervision, smart urban management has a strong demand for location service ability, basic spatial data acquisition and updating ability, and professional management data acquisition and updating ability. Therefore, BDS and remote sensing data is considered to be well applied in smart urban management.

The application of smart city management mainly includes the following contents.

(1) Based on the BDS compatible system, the urban management refined real scene information acquisition and processing system is used for real scene 3D

data acquisition and component data database construction within the scope of smart urban management supervision.

(2) Using the intelligent location service platform and its terminal device in the integrated application service platform of grid urban management, according to the needs of smart urban management, the user terminal can provide functions such as road inspection, supervision and evaluation, on-site law enforcement and law enforcement task processing, data analysis and supervision.

(3) The urban management special vehicle supervision system is used to implement intelligent management and control of sanitation vehicles, law enforcement vehicles and residue transport vehicles related to urban management.

(4) The urban management business dynamic supervision system based on high-resolution remote sensing and positioning real scene data fusion technology is used to carry out demonstration application in key management fields such as urban outdoor advertising, illegal construction, industry supervision, site residue, river supervision, post disaster analysis, etc.

The integrated application service platform of grid urban management satellite technology is divided into basic layer, data layer, platform layer, application layer and user layer.

The basic layer includes the network environment, hardware and software supporting facilities, site, security system, standard system, etc.

The data layer includes basic terrain data, satellite navigation and positioning data, satellite remote sensing image data, urban management component event and grid data, geocoding data, real image data and urban management business data.

The platform layer includes four core platforms: intelligent location service platform of urban management, basic application platform of grid urban management, integrated satellite technology application service platform of urban management and real scene information collection and processing platform, which provide support services for each functional module of the application layer.

The application layer includes information collection and processing, professional data update, special vehicle management, live image management, data statistical analysis, business dynamic supervision, urban management intelligent location service, etc.

The user layer is a functional system directly facing users, including urban management, government, supervision, command, processing and other systems based on satellite navigation and location services, as well as the supervision system for special urban management vehicles such as sanitation vehicles, waste vehicles and law enforcement vehicles.

Figure 6.7 shows the overall architecture of the grid urban management satellite technology integrated application service platform.

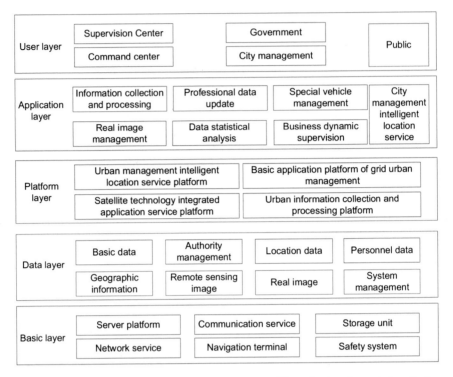

Fig. 6.7 Overall architecture of grid urban management satellite technology integrated application service platform

6.5 Application of Digital Construction

6.5.1 Introduction

In recent years, construction machinery industry of China has developed rapidly. According to the statistics of China Construction Machinery Association, by the end of 2016, the number of main products of construction machinery in China was 6.72–7.28 million. With the large-scale mechanization popularization of urban construction, domestic construction machinery vehicles are still growing at a fast speed. In addition to factories and workshops, the working environment of most construction machinery is more complex, harsh and changeable, such as rivers, mountains and forests. At the same time, with the continuous improvement of engineering quality requirements, the traditional operation mode of construction machinery has been difficult to meet the needs of high standard and high quality of modern engineering operation. It is urgent to promote the development of construction machinery to achieve new goals through the combination of external resources.

The new economic era in the twenty-first century is the era of intelligence. Intelligent technology has been gradually applied to various fields of social development. Construction machinery affects all aspects of the society and provides the necessary material conditions for people's life. With the progress of society, people pay more and more attention to the development process of construction machinery. The application of intelligence in construction machinery has become a scientific problem with research value.

In recent years, the product market related to construction machinery technology is monopolized by the United States, Russia and other developed countries, which makes the construction machinery technology make outstanding contributions in various fields of national development. Take the construction machinery products of the United States and Russia as an example. As one of the early products of Caterpillar Inc. in US, Caterpillar 992c loader is still serving in the coal mining industry all over the world with its tenacious vitality. The main reasons are as follows: ① the loader is equipped with wireless remote control system, which can realize its operation in dangerous areas through remote manual operation; ② the cab and roof protection with reinforcement support are provided safety guarantee for the operator. The biggest highlight of Russian unmanned mine dump truck is the unmanned driving and the accuracy of the operation process. The dump truck can go deep into the nearly kilometer tunnel for accurate operation, which improves the operation efficiency of construction machinery.

The research on intelligent technology of construction machinery in China began at the end of the twentieth century. The research is mainly carried out from four aspects: the roboticization of construction machinery, the intellectualization of excavator, the intellectualization of bulldozer and the technology of automatic tractor. The development of intelligent control system of domestic construction machinery mainly experienced four stages: introducing and learning foreign advanced technology, secondary development of core technology, imitation development and independent research and development. Around 2010, scientific research institutions in China began to design and develop the control strategy of construction machinery independently, mainly using the model-based development mode to carry out the systematic research and development of related construction machinery control system. At the same time, it was put into production in China, and the equipment intelligence was further improved. At present, Chinese intelligent construction machinery has entered a stage of rapid development. Based on the precise space-time service provided by BDS, we can solve various problems in the practical application of mechanical operation and form a technical architecture, which can provide the necessary reference for construction machinery to transform from a traditional industrial upgrading to a modern industry combined with electronic information, and to create the core competitiveness in the world market.

6.5.2 Application Scheme

The application of GNSS in the field of construction machinery integrates many cutting-edge technologies, such as BDS technology, communication technology, IOT, cloud computing technology, and so on. It is highly integrated to assist construction operations, providing information management and remote management for construction enterprises. Figure 6.8 shows the overall application framework of the guidance management system for precision construction of construction machinery based on BDS precision positioning.

1.   The software of BDS based construction machinery precise construction operation guidance management system

The system of guidance and management system for precise construction operation of construction machinery based on BDS precise positioning is shown in Fig. 6.9. Based on BDS precise positioning, the construction machinery precise construction operation guidance management system takes the intelligent monitor of embedded microprocessor as the core, adopts the expanded wireless communication module and BDS technology, as well as various technical means such as data acquisition, intelligent electronic monitoring, real-time control, geographic information system, information security, etc. All these methods can be interconnected according to the actual situation of the construction site. The high-precision integrated dual antenna mode is adopted to receive the differential correction data of the base station, determine and calculate the coordinates and posture of the front and rear antennas. The three-dimensional coordinates of the working parts can be accurately calculated combined with the sensors and the internal algorithm model of the construction

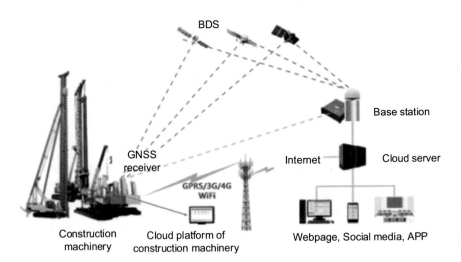

Fig. 6.8 Overall application framework of guidance management system for precise construction of construction machinery based on BDS

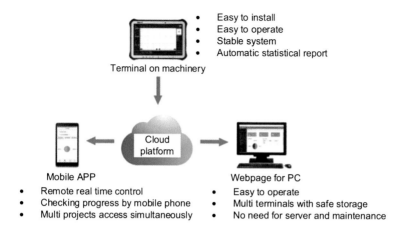

Fig. 6.9 Software system

machinery. The on-board controller can read three-dimensional electronic drawing data and compare with the positioning value, so as to give timely guidance to the operator in real time. The control and measurement accuracy can reach centimeter level, improve the work efficiency, and reduce the time consumption and error caused by measurement. At the same time, it can provide users with complete online cloud services such as creating construction tasks, operation management, construction progress monitoring, operation results statistics, automatic generation of construction records and as built drawings, etc. It can also provide a convenient and efficient project level data cloud platform for intelligent operation of construction machinery, which can grasp the quality and progress of the project anytime and anywhere, shorten the construction period and reduce the construction cost with greatly improvement of working efficiency.

The guidance and management system of construction machinery precision construction operation based on BDS precision positioning is the basic platform of IOT owned by construction machinery manufacturers. It establishes data exchange center, formulates unified standard communication protocol, and is compatible with terminal products of different manufacturers. Real time data are collected and storage in the local server of construction machinery manufacturers. The data are compatible with the existing system platform. Historical data are also retained. The system supports the smooth migration of existing GPS platform data, and supports data mining from the data uploaded from the terminal. Through data cleaning and mining, the location data (site change, regional distribution), workload data (working hours, fuel consumption) and working condition data (product related parameters) can be collected, counted and analyzed, and the report forms can be displayed, so as to realize intelligent search. It can realize remote control based on vehicle terminal, support accessing with DMS, ERP, CMR and other information systems of construction machinery manufacturers, and provide access interface for other satellite vehicle positioning systems to realize data sharing.

2. The hardware of BDS based construction machinery precise construction operation guidance management system

The guidance hardware terminal of construction machinery precision construction operation based on BDS precision positioning includes BDS reference station, BDS high precision receiver, multi-mode multi frequency antenna, industrial tablet computer and related sensors.

BDS high-precision receiver can provide heading and positioning information, based on inertial navigation technology, built-in high-precision satellite positioning module. Dual antenna design can greatly improve the space utilization. Combined with gyroscope, accelerometer and other sensors installed on the engineering vehicle, the position, heading angle, elevation angle, roll angle and other attitude information of the engineering vehicle are accurately output through the fusion algorithm. When the clearance condition is bad and the GNSS signal is blocked, the inertial navigation module can continuously output the accurate position information.

The multi-mode multi-frequency antenna adopts the multi-feed point design scheme to realize the coincidence of the phase center and the geometric center, so as to minimize the influence of the antenna on the measurement error. The antenna unit has high gain and wide beam pattern, which ensures the technical effect of low elevation signal and can work normally in some severe occlusion situations. The antenna is equipped with anti-multipath choke plate, which can effectively reduce the influence of multipath effect on measurement accuracy. At the same time, it adopts split design, waterproof and anti-ultraviolet, and can work normally in harsh environment.

The industrial grade tablet computer adopts a solid all aluminum structure, with long service life, high reliability, fast response, no tailing, no shaking, and ultra wide viewing angle. It can watch clear pictures from a large angle. At the same time, it integrates a resistive touch screen, and its panel can reach IP66 waterproof and dustproof level.

6.5.3 Application Function

Based on BDS precise positioning, the construction machinery precise construction operation guidance management system integrates the signal processing components and circuits of single function sensors such as tilt, depth, laser, etc. into one system with the help of microelectronics and micro processing technology, so as to reduce the size of the sensor, improve the reliability, and enhance the anti-interference ability. The system adopts open API architecture, object-oriented, interface oriented and domain driven design to abstract and design the system. It uses its own software platform based on Java to design and develop the system, with standard extension interface reserved to support the subsequent expansion of the system. After the system is deployed, it can not only support the functions of various mechanical products, but also provide different services for various users, such as providing vehicle value-added services for vehicle owners, providing data analysis services for depot leaders, and providing vehicle maintenance process management for after-sales personnel.

Based on BDS precise positioning, the construction machinery precise construction operation guidance and management system supports a variety of types of construction machinery and equipment, including pile driver, excavator, bulldozer, grader, roller, paver, etc. It can build an efficient, accurate and intelligent auxiliary construction module to realize the guidance and management of construction site informatization auxiliary construction operation.

Construction machinery and equipment manufacturing industry is an important embodiment of the scientific and technological level of a country. It is a strategic industry to provide technical equipment for national economic construction, and is an important way for countries in the world to develop their economy and improve their comprehensive competitiveness. But for a long time, the situation of safe operation of construction machinery is very serious. Because of the complex system composition and bad working environment, construction machinery products usually require high load and long-time operation. Major accidents such as machine damage and human death occur from time to time, which seriously affect the progress and benefits of construction projects and property safety. By giving full play to the core technical advantages of BDS precision positioning service, the application of BDS precision positioning in the guidance and management of construction machinery precision construction solves the technical problems of construction site safety supervision. The disorder, rework and schedule delay of on-site management are greatly reduced. The manpower, material resources and time cost savings for on-site construction are expected to reduce about 5–10%, which can effectively meet the needs of high efficiency and fine management of engineering construction. This effectively improves the accuracy and safety of engineering construction, provides effective reference for improving the independent innovation ability and core competitiveness of construction machinery and equipment manufacturing industry of China. The progress promotes the transformation of engineering site construction mode from relying on artificial subjective judgment to making decisions based on objective data. It is an important part of intelligent operation management of construction machinery and equipment.

6.6 Application in Water Field

6.6.1 Introduction

Water is the source of human life. However, with the city development, water pollution is becoming more and more serious. Water resources supervision and governance has become a major problem of urban development. Untimely water quality monitoring and flood warning are directly related to people's livelihood. In addition to water resources, water management also includes different links in the process of water resources development, utilization and protection, such as flood control and drainage, water supply, water use, drainage, water saving, sewage treatment, reclaimed water use, water environment protection and water ecological construction. With the increasing scale of water operation enterprises, how to improve business management level, effectively reduce operating costs, and how to effectively supervise the market-oriented operation of water plants by government departments are urgent problems to be solved. As far as water operation enterprises are concerned, the number of water plants to be managed is increasing year by year and they distributed all over the country. Enterprises need to realize real-time operation and management of water plants under their jurisdiction. As far as the government regulatory system is concerned, regulators do not directly intervene in the production management of operating enterprises, but must be in place to ensure the quality of water resources, a public product. If we only rely on limited human and material resources, it is very difficult to dynamically and comprehensively monitor the production status of many operating enterprises. Therefore, making full use of the achievements of modern science and technology development and information means is an inevitable choice to promote and drive the modernization of water affairs, enhance the social management and public service capacity of water industry, and ensure the sustainable development of water affairs.

Under the guidance of the concept of "smart water", the management of water group has gradually changed. They are based on GNSS technology, combined with data acquisition, transmission and other sensor equipment to detect the operation status of water system online. The visual way is applied to organically integrate the facilities of water management department to form the "water IOT". Through the digital management platform of water affairs, the water affairs group timely analyzes and processes massive data. It installs data acquisition front-end processor or data acquisition DSP module in each water plant and pump station, and transmits the production and operation data of automatic control system to the headquarters of water affairs group through wired/wireless network in real time for centralized storage and application. Through real-time monitoring and intelligent analysis of all kinds of key data, classification and grading early warning are provided. SMS, light, alarm sound are used to inform relevant responsible persons. Corresponding processing results are given to assist decision-making suggestions, so as to manage the whole production. Management and service process of water business operation system make it a more precise and dynamic way to achieve the state of "wisdom".

6.6.2 Application Scheme

The IOT, BDS, remote sensing, radar telemetry, video monitoring and information perception and other technical methods are fully applied to form a three-dimensional monitoring network for water safety, water resources, water ecology and water environment. An air space integrated water basic information collection and transmission system is built with reasonable layout, complete structure, complete functions and high sharing, so as to comprehensively improve the water quality. It also has the ability to monitor flood control and drainage management, water resources regulation and water environment improvement. The overall framework of smart water system is shown in Fig. 6.10. The framework is divided into three layers: IOT perception layer, data center layer and business application layer.

Fully combined with the business process, taking the unified operation of water information as the premise and basis, an integrated management system application

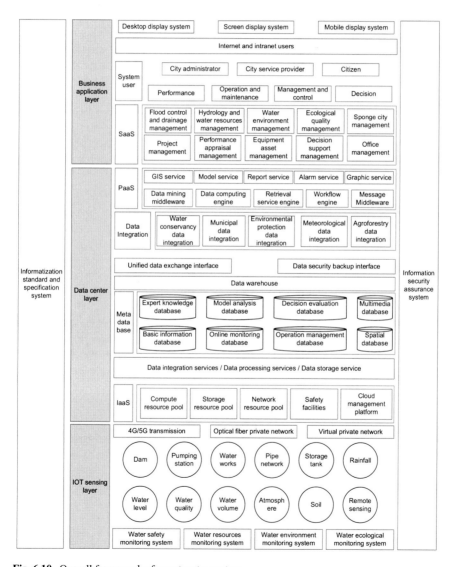

Fig. 6.10 Overall framework of smart water system

platform is established. Adopting the "Cloud + End" system construction mode, an integrated intelligent water application platform is built to adapt to management innovation and continuous optimization of information system. "Cloud" is a unified private cloud platform, including infrastructure as a services (IaaS), platform as a services (PaaS) and software as a services (SaaS), fully meeting the infrastructure services and software application services needs of water group and its subordinate units. "End" is the application end. Water group and its subordinate units are the application end of the private cloud platform, and the private cloud platform provides services for the application end. Smart water system comprehensively displays a map of pipe network safety, a map of pipe network leakage, a map of user service, sewage and other thematic maps. The system realizes auxiliary comprehensive operation analysis and decision-making of pipe network operation monitoring, DMA analysis, hydraulic model, emergency dispatch, customer service and secondary water supply based on various integrated data. The system integrates real-time dispatching technology with international advanced level to realize real-time model analysis and dispatching decision support, so as to help the water company build an advanced integrated monitoring, early warning, decision-making and command platform for urban water supply. Based on the construction of data center platform to break through the barriers of enterprise production data, various real-time production information of enterprises are comprehensively integrated. Based on GIS map display module to realize data integration, a dynamic information application platform of deep integration is built.

1. The perception layer of IOT

The IOT sensing layer uses various sensors, such as water turbidity sensor, water level sensor and water pressure sensor, to obtain all kinds of information of the device. Data are collected through the device console or special control box (PLC), and information obtained through the sensing layer is transmitted and processed through various networks (including Internet, WiFi, broadband and telecommunication network). The platform can access all kinds of sensors on the market, instead of binding its own exclusive hardware, so as to give users more choice space, maximize the use of existing acquisition hardware, and reduce the implementation cost of the system.

2. Data center layer

Data center layer is the data information infrastructure and important support of smart water business expansion, which undertakes the tasks of data storage, data cloud node, cloud data sharing, cloud computing services, cloud computing analysis, remote disaster backup services, etc. The collected and stored data include all kinds of BDS navigation/positioning data, water condition monitoring data, water management data, emergency dispatching data, etc. Through cloud computing technology, it can deal with massive data concurrent access and massive heterogeneous data query, and realize data interaction and sharing of various application systems.

3. Business application layer

Business application layer is the interface between IOT and users. It combines with industry demand to realize intelligent application of IOT. Automatic alarm for abnormal state, such as limit low water level alarm and interlock protection, high water level alarm, etc. The cameras can monitor the status of equipment in real time from multiple angles, especially in remote areas and places where people are not easy to visit (including reservoirs, sewage wells, narrow spaces, etc.).

6.6.3 Application Function

Smart water system can provide a solid and reliable data sharing platform for each water group, and provide perfect geospatial data services for each department. It not only solves the problem of dynamic update of pipe network data, but also saves the funds invested by each department in basic geographic and thematic data collection. Closely combined with the business process of water supply management, the user departments can share all kinds of data, timely and effectively provide accurate and reliable data, and can synchronously consult data anytime and anywhere on the large screen, PC and mobile terminals. The system can realize all kinds of business subject analysis, data filling and so on, greatly reduce the workload of statistical personnel in all departments, and abolish manual statistics and paper reports by automatically extracting reports. Managers at all levels can timely grasp the production and operation status to solve the problems caused by incomplete data collection or delayed system response in the past. The ability of data-based decision-making is improved to ensure the safety of water supply, and conduct dynamic and visual simulation and prediction of sudden pollution events. The decision-making support for accident emergency treatment provides reference for planning, design, scheduling and reconstruction and expansion. In order to improve the operation efficiency and management level of water supply enterprises, the reasonable reference basis is provided for various applications such as the scheme.

In the aspect of information sharing, the information sharing service desk is built to realize the release and sharing of spatial basic geographic information, pipe network information, business data and monitoring data among various departments, enterprises and enterprises, enterprises and governments. The sharing provides basis for authority management, data update and security management.

6.7 Application of Ecological Environment Monitoring

6.7.1 Introduction

BDS spatiotemporal ecological environment housekeeper service platform is an important starting point for the local government to solve problems and help the

government taking the major responsibility of ecological civilization construction. The platform can further promote the prevention and control of air, water and soil pollution, comprehensively implement various strengthening measures, help governments at all levels correctly handle the relationship between economic development and environmental protection with advanced science and technology and more perfect solutions, improve the efficiency of government funds, and realize the benign development of local government ecological environment. Figure 6.11 shows the BDS spatiotemporal ecological environment housekeeper service platform.

In order to deeply solve the environmental quality problems existing in various regions, it provides sub services including top-level design, construction of ecological monitoring network, analysis of pollution causes, formulation of management and control countermeasures, economic impact analysis, preparation of up to standard planning and effect simulation evaluation, etc. Relying on expert consultation and big data analysis technology of ecological network, BDS spatiotemporal ecological environment manager service platform assists government decision-making. In order to improve the effectiveness and economy of decision-making, we should guide the upgrading and transformation of key industries quickly, accurately and targeted, and carry out inverse simulation and evaluation of the effect, so as to achieve a win-win situation of pollution prevention and economic development through effective governance and supply side reform. Figure 6.12 shows an example of air pollution monitoring technology of BDS spatiotemporal ecological environment manager ecological monitoring network. Figure 6.13 shows the sub service platform of BDS spatiotemporal ecological environment manager.

6.7.2 Application Scheme

1. Construction of unified management and control platform

The existing system cannot meet the needs of modern office and collaborative linkage of environmental protection administrations. In order to solve the pain point of "data island" of environmental protection, realize data interconnection and achieve the purpose of comprehensive display and analysis of environmental related information,

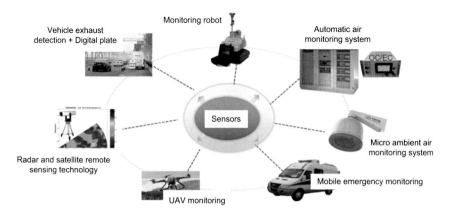

Fig. 6.12 Air pollution monitoring technology of BDS spatiotemporal ecological environment manager ecological monitoring network

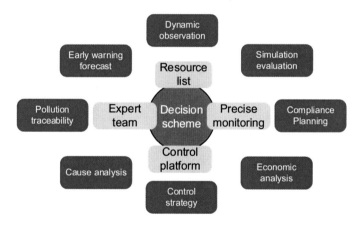

Fig. 6.13 Sub service platform of "Beidou spatiotemporal ecological environment manager"

the main construction contents of unified management and control platform are as follows:

(1) Integrating the existing systems inside and outside environmental protection.
(2) System data support platform and information sharing platform (environmental protection resource directory).
(3) Grid management system.
(4) Enterprise environmental information management system (one enterprise one file).
(5) Environmental geographic information thematic map application system.
(6) Information comprehensive analysis and display system.
(7) Environmental total quantity control system.
(8) Collaborative OA system.

(9)   Intelligent environment integrated portal system.

(10)   Smart environment app.

The design of command center and "leader cockpit" is convenient for the unified command and dispatch of environmental protection work. As the headquarters for centralized handling of air quality problems, it can improve efficiency and reduce the consumption of human and material resources. The leader cockpit combined with business intelligence (BI) technology can realize the comprehensive display of data statistics. Real time data presentation and analysis results display, including pollution calendar, pollution ranking, diachronic data browsing, trend analysis, statistics and other data display and analysis functions, provide the current situation of environmental quality in the whole region to facilitate leaders at all levels quickly understand (see Figs. 6.14 and 6.15).

2.   Construction of ecological environment monitoring network

In order to comprehensively strengthen the construction of ecological environment monitoring network, it is necessary to supplement and improve the existing monitoring grid to form a set of comprehensive, scientific and effective ecological environment monitoring network system with the integration of heaven, earth and space. The content includes following aspects.

Fig. 6.14 Effect of large screen control center

Fig. 6.15 Display of leader cockpit interface

(1) Carrying out remote sensing monitoring of vehicle exhaust to improve the level of environmental protection supervision and management of in use vehicles;

(2) Carrying out the construction of automatic monitoring points of air quality in port terminals to promote the control of port ship pollution;

(3) Promoting the construction of VOCs online monitoring equipment to strengthen the monitoring and supervision of emission enterprises;

(4) TSP on-line automatic monitoring and video monitoring equipment will be installed at the construction site and the residual sludge recovery site, dust pollution control will be carried out in depth, and dust monitoring will be implemented on key roads;

(5) Installing on-line automatic monitoring equipment for lampblack emission in the catering industry at or above the benchmark stove in the core control area, and strengthen the governance and supervision of the catering industry;

(6) BDS UAV intelligent monitoring and remote sensing monitoring are adopted to improve the monitoring ability of key air pollution sources;

(7) In order to strengthen the construction of air monitoring capacity and air quality prediction and judgment work, we carried out the construction of air compound pollution super observation station and air quality monitoring substation, and carried out the construction of high-density grid monitoring network system (including monitoring TVOC, NO_x, O_3, etc.) in the core control area.

3. Environmental big data aided decision making and evaluation services

In view of the current situation of excessive ozone and other pollutants and emission reduction targets, BDS spatiotemporal ecological environment housekeeper project proposes targeted environmental protection big data decision-making and other services, and its flow chart is shown in Fig. 6.16.

Fig. 6.16 Flow chart of
environmental big data aided
decision making

(1) Monitoring sites: with professional technology and practical experience, the project provides design guidance for the construction of ecological monitoring network and the selection of monitoring types and sites, to ensure the effectiveness and reference of monitoring results.

(2) Analysis and diagnosis: on the basis of big data analysis of environmental protection, the project provides qualitative and quantitative identification of pollutant sources in environmental receptors which can effectively analyze the emission types and contribution rate of compound pollutant emission sources, and provide data basis for targeted treatment schemes.

(3) Decision making consultation: in the case of decision-making, analysis and judgment, and emergency response, it is not only necessary to unify the regulatory function of the management and control platform and the analysis of environmental protection big data based on models and algorithms, but also need to combine the experience judgment and analysis of environmental monitoring, governance experts, environmental protection bureau and other professional and technical personnel. Targeted conclusions are drawn to ensure the rapid implementation and expected effect.

(4) Effect evaluation: on the basis of the above comprehensive work, the project provides standard planning proposals for local governments, and evaluates the effect of the contents mentioned in the standard planning. The structure of effect evaluation can effectively guide the monitoring, analysis and decision-making of big data.

6.7.3 Operation Pattern

Project construction generally adopts EPC + O mode, which means that the contractor is responsible for the whole process of design, procurement, construction and installation of the project, and is responsible for the later operation service,

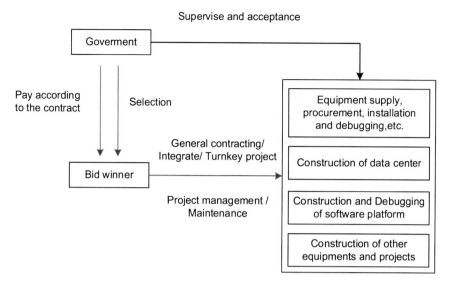

Fig. 6.17 Architecture of EPC + O mode

that is, the turnkey service mode that the contractor is responsible for the operation, as shown in Fig. 6.17.

6.8 Application of Air Pollution Supervision

6.8.1 Introduction

According to statistics, at present, the main air pollution sources can be divided to the following four categories: bulk coal combustion, chemical enterprise pollution, motor vehicle exhaust and dust. The above four are not single fixed pollution sources. Compared with single fixed pollution, the complex degree of monitoring and treatment of these compound pollution is higher, which puts forward higher requirements for the management and control technical means and decision-making implementation process of the regulatory department. In addition, from the experience of urban application, it is difficult to coordinate the cross-industry and administrations, such as mobile source pollution control, dust pollution control, bulk coal pollution control, comprehensive prevention and control. Due to the lack of information exchange and data sharing among various management organizations, it is usually impossible to form an effective joint force, which eventually leads to the ineffective implementation of policies and the difficulty in achieving the goal of quality improvement.

The multi-level and wide coverage of the integrated air environment monitoring system, which integrates the advantages of advanced technology and equipment, makes full use of all-weather, multi-scale and multi category monitoring means, improves the environmental monitoring ability. At the same time, the mode is changed from monitoring to the whole process supervision of pollution discharge, which effectively controls the illegal discharge of pollutants and gradually controls and reduces the total amount of pollutants discharged to the environment. The ability of environmental supervision is improved.

The BDS based atmospheric monitoring application scheme will establish a grid environmental monitoring system at the provincial, municipal, district, town and village levels in accordance with the requirements of "determining regions, responsibilities, personnel, tasks and assessment". This scheme can strengthen the leading responsibility of governments at all levels for environmental supervision and law enforcement in their own administrative regions, and assign supervision and law enforcement personnel according to regions, so as to implement the supervision responsibility to units. Combined with grid monitoring technology, we can rank and assess grids at all levels, and screen out hot grids to focus on. The interconnection of business systems within environmental protection departments is achieved to break the "data island". A data sharing and business collaboration platform for regional atmospheric environmental problems is formed. This platform realizes the unified management of business systems, unified storage of data resources, and unified planning of business processes. Real-time sharing of relevant data with other government departments breaks the data barrier of cross department environmental protection related business. This "big environmental protection" promotes the continuous improvement of daily management and quantitative assessment level of atmospheric environment.

6.8.2 Application Scheme

1. The intelligent grid management system of atmospheric environment monitoring based on BDS

The BDS based intelligent grid management system for integrated atmospheric environment monitoring takes satellite remote sensing technology, advanced LiDAR monitoring technology and BDS ground grid monitoring technology as the core. The system adopts multi-level and multi-scale network monitoring scheme, and establishes BDS integrated three-dimensional atmospheric environment monitoring system and intelligent inspection network based on BDS. In order to comprehensively control the air pollution situation in the monitoring area and help the environmental managers to manage scientifically and implement comprehensive policies, a standardized inspection management platform is established.

Firstly, satellite remote sensing and atmospheric model monitoring are established to monitor the distribution and overall trend of air pollutants in a large scale. Secondly, LiDAR remote sensing network monitoring system is established to locate and track the air pollution sources. Finally, through the establishment of densely distributed

near ground monitoring grid stations, the accurate analysis of the causes of air pollution sources is realized. The "three-dimensional, multi-means, high-density, wide coverage" BDS air pollution monitoring and management mode is formed to obtain the local and surrounding multi-scale, multi parameter air pollutant distribution information, and to master the temporal and spatial variation law of pollutants (Fig. 6.18)
.

2.   The whole area coverage monitoring by satellite remote sensing

Through the data processing of multi-source and multi temporal atmospheric environment and pollutant monitoring satellites, the system provides important technical support for the trend and source judgment of air pollution.

The air environment and pollutant monitoring satellite has the ability of active and passive air environment comprehensive monitoring, which can provide data support for air environment pollution monitoring and early warning after launch.

The main task of the satellite is to detect the following environmental elements:

(1)   Fine particles (PM2.5, etc.);
(2) CO_2;
(3)   Pollution gases (SO_2, NO_2, O_3, etc.);
(4)   Incineration of straw and others;
(5)   Dust weather in the north.

Satellite data cover a wide range. It can be used to judge the trend of air pollution. However, because the satellite has a transit period, it is difficult to meet the time and spatial resolution requirements of regional air pollution monitoring. Therefore, it is necessary to cooperate with the ground high-precision spatiotemporal monitoring means to achieve high-precision real-time monitoring of regional air pollution.

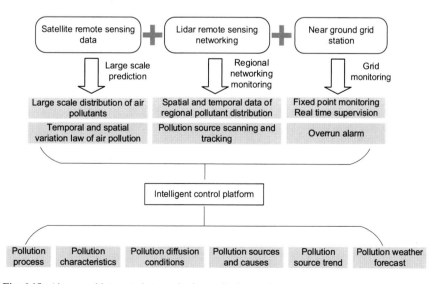

Fig. 6.18   Air ground integrated atmospheric monitoring system

3.   Advanced LiDAR air pollution monitoring network

Many world-renowned environmental and atmospheric experts have published articles in top journals pointing out that aerosol vertical distribution is the most important parameter for atmospheric transport, air quality change research and climate application. Because of its short wavelength, LiDAR can directly interact with aerosol or cloud particles. At the same time, it has the outstanding characteristics of high monochromaticity, high directivity and high vertical resolution. It is recognized as the best means of aerosol vertical observation. It can be used to measure the changes of aerosol, cloud, visibility, atmospheric composition, air wind field, atmospheric density, temperature and humidity. The diffusion process of environmental pollutants over the city can be monitored effectively.

High precision, multi-element and real-time atmospheric environmental monitoring is the main basis of atmospheric environmental governance in China. It is very important to build a new type of remote sensing monitoring equipment for air pollution and realize the innovation of atmospheric detection instruments in the industry. Atmospheric LiDAR can be used to observe the distribution and time evolution of smoke and dust in horizontal direction over the city. The system has outstanding advantages of small volume, easy to move and carry, and can automatically observe in all-weather. At the same time, its modular structure ensures the stability of LiDAR operation and the reliability of detection data. These advantages make the observation cost of LiDAR network lower and LiDAR application more extensive.

LiDAR remote sensing algorithm inversion can provide more accurate and reliable tracking and positioning of pollution sources in the laser scanning monitoring of key areas.

The operation method of LiDAR atmospheric monitoring is as follows (see Fig. 6.19)

(1)   The concentration of particles and aerosols within a radius of 10 km and 24 h continuously monitor.
(2)   Two dimensional particle pollution map was obtained by horizontal scanning.
(3)   Accurate longitude and latitude of the most concentrated area of pollutants.

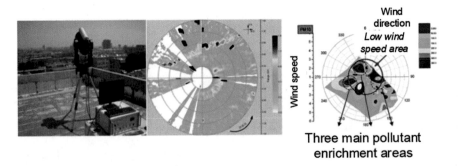

Fig. 6.19 LiDAR atmospheric monitoring technology

(4)   It is used for height monitoring and forms three-dimensional monitoring results with ground micro air monitoring station.

By comparing the extinction images of different wavelengths of LiDAR and combining with the near surface meteorological data, the optical characteristics of different pollution sources can be compared. The pollution characteristics of different regional pollution sources can be mastered. Furthermore, the local pollution removal mechanism can be deeply analyzed by combining the radar images with the macro weather situation.

Because it is difficult for a single LiDAR to determine the source of pollutants, in order to better analyze the local pollution situation, it is necessary to place multiple LiDARs on the main transportation channel and the location with characteristic terrain. The local pollution situation and the source of pollutants can be monitored and analyzed.

By inputting the output results of the meteorological model into the atmospheric model, and adjusting the model parameters locally, we can track the direction of heavy pollution, analyze the pollutant diffusion trajectory within 72 h after the occurrence of external pollution, and display it dynamically on the GIS map, which provides assistance for the formulation of pollution emergency measures. When the LiDAR finds the external pollution, it can combine the backward trajectory to judge the external pollution source, and combine the satellite data to explain the regional pollution qualitatively and quantitatively. LiDAR is the only way to detect the distribution of pollutants in the vertical direction by remote sensing. Therefore, LiDAR is used to perform longitudinal observation, draw the vertical structure of pollutants, study the change of height distribution of pollutants, and assist the backward trajectory model to complete the tracking and exploration of pollutants. At the same time, longitudinal observation can quantitatively describe the height of atmospheric boundary layer. It is an important means to forecast severe pollution weather.

Through the network observation of multiple stations, the output results of the meteorological model are input into the atmospheric model. The model parameters are adjusted locally to realize the tracking of heavy pollution. When the LiDAR finds the external pollution, it can combine the backward trajectory to judge the external pollution source, and combine the satellite data to explain the regional pollution.

The technology has the following characteristics

(1)   Atmospheric particulate matter network observation: LiDAR network, regional pollution event tracking.
(2)   Aerosol vertical distribution observation: aerosol, cloud three-dimensional distribution and time change monitoring.
(3)   Particle movement observation: pollution source scanning and tracking to capture the location of pollution sources.
(4)   Without contact measurement, we can obtain the observation data of the area which is not easy to reach.

Build a multi-site LiDAR remote sensing network system with a laser horizontal scanning radius of 5 km. The areas that cannot be covered are covered by satellite data to achieve full coverage, real-time data of key areas and quasi real-time data of other areas. To track the direction of heavy pollution, it can analyze the diffusion track of pollutants within 72 h after the occurrence of external pollution, and dynamically display it on the GIS map, providing assistance for the formulation of pollution emergency measures.

The system can provide accurate regional air pollution daily report (heavy pollution days), weekly report, monthly report, annual report, and pollution situation analysis report. Proposals, plans and remediation strategies can be supported by the professional data for the government and authorities at all levels.

4. Ground grid monitoring

Through the advanced LiDAR air pollution remote sensing network monitoring method, we can judge the situation of the key pollution sources in the monitoring area, but the scale is still relatively large. We need to combine the ground BDS grid monitoring method to accurately locate the pollution sources and improve the distribution map of the main pollution sources.

The full coverage and three-dimensional air quality monitoring and scanning system is an important part of BDS intelligent environment system. The system aims at urban residential areas, rural towns, industrial parks, key industrial enterprises, road traffic, construction sites, regional boundaries, pollutant transmission channels and other environmental monitoring objects, forming an online monitoring platform covering the whole region. It can not only monitor the dynamic changes of major pollutants in the region in real time, and quickly capture the abnormal emission behavior of pollution sources and give real-time warning. Moreover, it can analyze the main sources of regional pollution through big data, achieve more accurate early warning and prediction, achieve targeted treatment of regional pollution, and effectively screen cross-border pollution.

1) Micro ambient air detection station

The measurement parameters of micro ambient air detection station are mainly as follows.

(1) Particulate matter: PM2.5, PM10.
(2) Conventional pollution sources: SO_2, NO_2, CO, O_3.
(3) Weather (optional): temperature, humidity, wind speed, wind direction, air pressure.
(4) Environment (optional): noise, electronic nose, video monitoring.
(5) Characteristic pollution sources (optional): VOCs, H_2S, HCl, HCN, NH_3, etc.

2) Distribution principle

Firstly, the standard stations are set up according to the actual situation. In order to analyze the pollution situation more pertinently, the scheme of grid combined with intensive monitoring of key pollution sources is adopted. In order to evaluate the contribution of various types of pollution sources to the standard station, the grid distribution is carried out within 5 km of the standard station according to the principle of 1 km × 1 km. The intensive monitoring is carried out at the key emission enterprises, roads, restaurants and other key sources around the standard station.

In order to better evaluate the pollution sources, meteorological monitoring equipment is set up near the standard station, which can be used to explain the high value events of the street town evaluation point. The equipment evaluates the impact of the surrounding pollution sources on the standard station, and can be used to evaluate the impact of external pollution transmission on the region.

At the same time, the distribution of micro stations should follow the following principles.

(1) Representativeness: the basic grid layout should be representative, which can objectively reflect the ambient air quality level and change law in a certain space, and objectively evaluate the regional ambient air condition.

(2) Comparability: the environmental conditions of the same type of monitoring points should be consistent as far as possible to make the data obtained by each monitoring point comparable.

(3) Scientificity: each grid of the ambient air quality grid monitoring system should consider the comprehensive environmental factors such as natural geography and meteorology in the distribution area, as well as the social and economic characteristics such as industrial layout and population distribution. The layout should reflect the pollution status and change trend of the main functional areas and main air pollution sources in the region.

(4) Rationality: the rationality of grid distribution is mainly reflected in different distribution methods for different pollution subjects, and appropriate encryption for key areas or enterprises.

3) Distribution of air quality micro stations

Air quality micro station is the main equipment specially used for high-density grid, fine mapping of air pollution, positioning and monitoring of pollution sources.

4) Meteorological monitoring

The meteorological monitoring equipment can grasp the meteorological conditions of each area in real time, and can cooperate with the monitoring data of micro station to monitor the pollution transmission process in real time. It traces the source accurately and predicts the pollution transmission path.

As shown in Fig. 6.20, the five-parameter meteorological product is displayed.

5) Mobile monitoring

By using social resources, carrying air quality sensing devices on vehicles, and integrating multi-dimensional big data such as urban road monitoring system, traffic flow, population density and urban functional areas, we can achieve "multi network integration" to better serve urban development, environmental governance, government planning and other aspects. It can accelerate the construction process of smart

Fig. 6.20 The
five-parameter
meteorological product

city, and timely discover the problems. At the same time, based on the accurate positioning of BDS, the pollutant distribution map of the whole city can be drawn, which provides a scientific and efficient management means for air pollution control.

The implementation of the air environment monitoring technology based on the integration of remote sensing and BDS can promote the technological of air pollution monitoring to the top level in China. The system can be built as the leading air pollution monitoring demonstration system in China. It realizes the accurate monitoring of key monitoring areas in the area, and finally forms a "three-dimensional, multi-means, high-density, wide coverage" BDS based air pollution monitoring and management mode.

5.   The intelligent grid management system of air environment monitoring based on BDS

The intelligent grid management system of air environment monitoring based on the integration of BDS adopts the fusion technology of BDS and multi-dimensional GIS, which displays the distribution of pollution sources, real-time monitoring of environmental quality, and changes of air pollution trend on a map. The system truly realizes "front perception with IOT, temporal analysis application, virtual simulation management, and multi-dimensional GIS analysis". The integrated GIS visualization application innovation mode realizes the intuitive control of environmental quality and management, and strongly supports the environmental protection decision-making and supervision.

The system will integrate satellite remote sensing monitoring data, LiDAR network monitoring data, BDS ground grid monitoring data, standard site data, meteorological data, forecast data, etc. It can realize the functions of pollution event alarm, pollution source analysis, regional ranking, air quality forecast, etc. The system also effectively combines the integrated monitoring data with the standard site monitoring data, analyzes and formulates daily, monthly, quarterly, semi annual, annual and special reports of air quality. The main regional environmental problems are analyzed. The governance direction is clarified. The system provides simple, practical and effective air quality standard management tools for urban environmental managers. Figure 6.21 shows the air pollution prevention and control grid monitoring command center.

6. Real time map of monitoring point

All the monitoring sites in the system are classified and displayed according to their administrative regions. The icon color of the monitoring site is dynamically displayed according to the current air quality index AQI. The specific geographical location is indicated above the icon, which is convenient for users to intuitively grasp the deployment and air environment quality status of the monitoring sites in each administrative region. The system provides a variety of map effects to display the air quality of the location of the sub station in real time and other real-time data.

7. Real time view of site data

After users click the monitoring point icon, the system will automatically display the information of AQI, site location, primary pollutants, releasing time, real-time data of various monitoring factors, etc. The AQI value is displayed with the color to visually display the current pollution situation of the site. The monitoring factors can be customized according to different needs. The display time is divided into real-time status value, the latest one hour value, the latest 24 h value, etc.

8. Remote video real-time monitoring of site environment

Video monitoring equipment can be installed on the monitoring site to intuitively understand the surrounding conditions of the monitoring site and the real-time emission data of pollutants through the window view. When the concentration of the

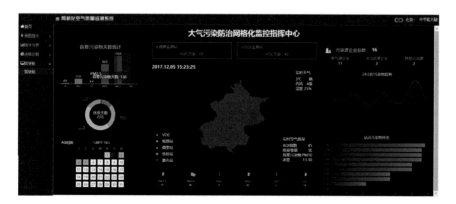

Fig. 6.21 Air pollution prevention and control grid monitoring command center

surrounding pollution sources exceeds the standard, it can be automatically captured
to provide the basis for the supervision and law enforcement of the public and environmental protection departments, and understand the real-time situation of the monitoring equipment. When the data is abnormal, the scene situation can be judged by
returning the image data. When force majeure occurs, the details of the accident can
also be judged according to the image data. Figure 6.22 shows the returned image
data.

9.   Early warning and daily notification

The system provides early warning and daily report notification functions, including
over limit early warning, disconnection early warning and abnormal value early
warning. When the monitoring value exceeds the standard, the data connection
is interrupted. When abnormal value occurs, the system will automatically send
reminder to the contact to ensure the normal and stable operation of the system.
The daily report notification will send the daily average air quality index of each
administrative region within its jurisdiction to the person in charge. Therefore, the
environmental manager knows the change of ambient air quality in time and knows
the detailed information at the first time when the air quality deteriorates.

10.   Data chart

The data chart supports line chart, bar chart, table and other forms. The display
contents include air quality index, minute value and hour value of the concentration
of various monitoring factors. It is convenient for users to view the change trend
of air quality and the change of pollutant concentration in a period of time. At the
same time, it can be used for comparative analysis of various parameters between
monitoring sites. Users can set the display time interval. Jpg, PDF, Excel, Word
and other formats are supported for data export and printing. Figure 6.23 shows
the comparative analysis of air quality composite index, and Fig. 6.24 shows the

Fig. 6.22 Returned image data

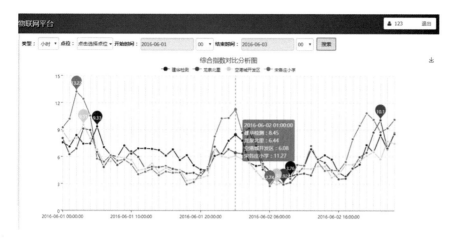

Fig. 6.23   Comparative analysis of air quality composite index

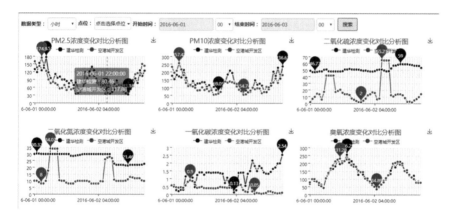

Fig. 6.24   Comparative analysis of pollutant concentration changes

comparative analysis of pollutant concentration change.

11.   Environmental quality ranking

According to the personalized needs of relevant environmental management departments and users, the system sets up an independent ranking system. At present, AQI
(air quality index) is used to provide daily ranking and hourly ranking data. Users
can query the ranking information and historical data of the day. In addition to AQI,
it also lists the hourly value, daily average value and primary pollutants of PM10,
PM2.5, CO and other monitoring factors.

12.   Automatic generation of AQI real time report and daily report

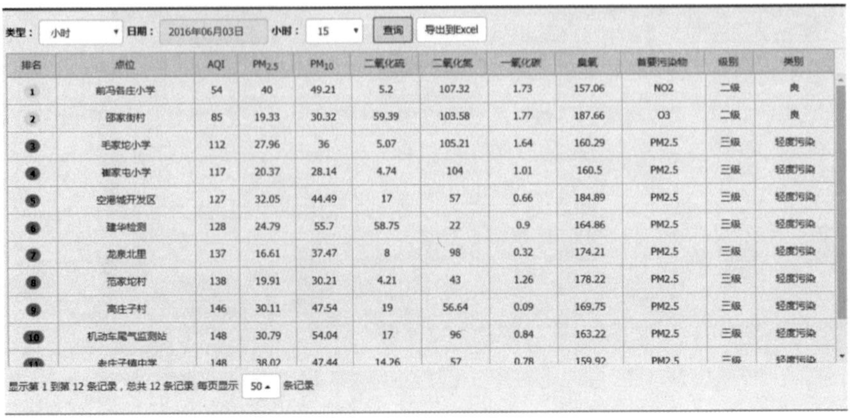

类型： 小时 日期： 2016年06月03日 小时： 15 查询 导出到Excel

排名	点位	AQI	$PM_{2.5}$	PM_{10}	二氧化硫	二氧化氮	一氧化碳	臭氧	首要污染物	级别	类别
1	前冯各庄小学	54	40	49.21	5.2	107.32	1.73	157.06	NO2	二级	良
2	邵家街村	85	19.33	30.32	59.39	103.58	1.77	187.66	O3	二级	良
3	毛家坨小学	112	27.96	36	5.07	105.21	1.64	160.29	PM2.5	三级	轻度污染
4	崔家屯小学	117	20.37	28.14	4.74	104	1.01	160.5	PM2.5	三级	轻度污染
5	空港城开发区	127	32.05	44.49	17	57	0.66	184.89	PM2.5	三级	轻度污染
6	建华检测	128	24.79	55.7	58.75	22	0.9	164.86	PM2.5	三级	轻度污染
7	龙泉北里	137	16.61	37.47	8	98	0.32	174.21	PM2.5	三级	轻度污染
8	范家坨村	138	19.91	30.21	4.21	43	1.26	178.22	PM2.5	三级	轻度污染
9	高庄子村	146	30.11	47.54	19	56.64	0.09	169.75	PM2.5	三级	轻度污染
10	机动车尾气监测站	148	30.79	54.04	17	96	0.84	163.22	PM2.5	三级	轻度污染
11	老庄子镇中学	148	38.02	47.44	14.26	57	0.78	159.92	PM2.5	三级	轻度污染

显示第 1 到第 12 条记录，总共 12 条记录 每页显示 50 条记录

Fig. 6.25 Air quality index ranking

According to the technical requirements of ambient air quality index (AQI), the system can automatically generate real time and daily data reports. The released indicators include the monitoring site information, AQI, primary pollutants, air quality index category and air quality index description of each monitoring site. It can automatically generate multiple report formats such as Word, Excel and PDF, which can be used for air quality control. The quality index ranking is shown in Fig. 6.25, and the air quality data report is shown in Fig. 6.26.

13. Source analysis of pollutants

After collecting the data from every site, the platform calculates and analyzes the statistical values of various pollutants, and initially establishes the point pollution source model, while the current method is the pie chart analytical method of the proportion of the primary pollutants. If the monitoring point conditions permit on-site sampling, the pollutant comparative analysis can be carried out more accurately. The pollutant proportion model of each time period can be combined with the current situation of the region to analyze the specific pollution sources and the actual situation of the site, and provide targeted treatment plan (see Fig. 6.27).

14. Equipment monitoring

The system can achieve real-time monitoring to the online monitoring instrument and the upload data. The operation status and progress of the equipment is obtained. When the current data acquisition equipment fails, the system will automatically provide alarm to facilitate the person in charge of the site to know in time, and take corresponding measures to ensure the normal and stable operation of the system.

15. Dynamic cloud image display of environmental data

Due to the difference of air quality between regions, the system renders the difference in the form of pollutant concentration cloud image in real time based on the

日期: `2016年06月05日`   小时: `15` ▼   [查询] [导出到Excel]

城市名称	监测点名称	二氧化硫（SO₂）1小时平均		二氧化氮（NO₂）1小时平均		颗粒物（粒径小于等于10μm）1小时平均	颗粒物（粒径小于等于10μm）24小时平均		一氧化碳（CO）24
		浓度/（μg/m³）	分指数	浓度/（μg/m³）	分指数	浓度/（μg/m³）	浓度/（μg/m³）	分指数	浓度/（mg/m³）
唐山市	唐山对外经济贸易学院	10.37	3	42.37	21	79.58	108.4	108	0.55
唐山市	老庄子镇中学	14.1	5	25.97	13	126.13	135.3	93	0.96
唐山市	范家坨村	29.3	10	30.63	15	123.5	155.09	103	0.44
唐山市	机动车尾气监测站	18.43	6	41	20	132.7	171.86	111	0.93
唐山市	高庄子村	17.8	6	24	12	59.89	110.53	80	0.95
唐山市	邵家街村	14.65	5	78.07	39	79.57	99.53	75	0.44

显示第 1 到第 13 条记录，总共 13 条记录 每页显示 25 ▲ 条记录

日期: `2016年06月03日`   [查询] [导出到Excel]

时间：2016年06月03日

城市名称	监测点名称	二氧化硫（SO₂）24小时平均		二氧化氮（NO₂）24小时平均		颗粒物（粒径小于等于10μm）24小时平均		一氧化碳（CO）24小时平均		臭氧（O₃最大1小时平均		臭氧
		浓度/（μg/m³）	分指数	浓度/（μg/m³）	分指数	浓度/（μg/m³）	分指数	浓度/（mg/m³）	分指数	浓度/（μg/m³）	分指数	浓度
唐山市	唐山对外经济贸易学院	12.99	4	79.58	99	171.61	111	1.14	28	185.22	82	
唐山市	龙泉北里	10.01	3	79.96	100	138.66	94	0.77	19	174.21	68	
唐山市	机动车尾气监测站	12.47	4	79.32	99	169.16	110	0.9	22	186.8	84	
唐山市	空港城开发区	12.5	4	47	59	102.25	76	0.71	18	184.89	81	
唐山市	老庄子镇	13.63	5	46.66	58	115.45	83	0.78	20	176.54	71	

显示第 1 到第 12 条记录，总共 12 条记录 每页显示 25 ▲ 条记录

Fig. 6.26   Air quality data report

类型: `小时` ▼   点位编号: `建华检` ▼   时间: `2016-06-03`   `00` ▼   [查询]

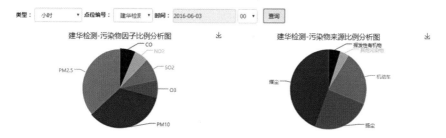

Fig. 6.27   Pollution source analysis

monitoring values in each region. The cloud image takes the hourly value of each site, and uses different colors to represent the air quality index AQI, so as to achieve the comprehensive display of air quality in a large range from "point" to "surface".

16.   Air quality and meteorological data export

The system provides the function of exporting air quality and meteorological data. The user can export the data after setting the time type, site and time period, including site information, data update time, concentration value of six conventional parameters, main pollutants and AQI. Among them, the data efficiency is calculated according to the national standard. The minute value is subject to the later data transmission judgment. The hour value is subject to the 45 min value collected every hour. The daily average value is subject to the 22 h value collected every day. The other time intervals are subject to the daily average effective days. Figure 6.28 shows an example of device monitoring data.

17.   Site management

In the site management module, the basic operations such as adding, modifying, checking and deleting the monitoring site information can be realized. Site information includes monitoring point name, address, longitude and latitude, site ID, area name, etc. The site management module can also realize the dynamic management of the site information (the area and number are in the locked state), and can configure the name, longitude and latitude, ranking, publicity, offline warning and other options (see Figs. 6.29 and 6.30).

18.   SMS configuration

In the SMS configuration module, we can view the SMS configuration details. In the item addition module, we can add SMS push personnel information and sending content. In the option editing module, we can manage the push content of the receiving SMS users. The configured information content includes warning information, daily report, status warning and offline warning. After setting, the personnel in the list can receive SMS information, as shown in Fig. 6.31.

19.   Early warning of pollutant concentration

点位编号	点位名称	在线状态	最新上线时间	最新注册时间	最新上报时间
XQ003	建华检测	✔	2016-06-19 09:40:18	2016-06-28 15:20:23	2016-07-15 13:55:01
XQ004	唐山对外经济贸易学院	✔	2016-06-06 17:01:38	2016-06-28 15:20:47	2016-07-15 13:55:38
XQ005	龙景北里	✔	2016-06-19 09:26:50	2016-06-28 15:20:03	2016-07-15 13:55:27
XQ006	机动车尾气监测站	✔	2016-06-28 05:06:11	2016-06-28 15:20:18	2016-07-15 13:55:01
XQ007	空港城开发区	✔	2016-06-23 12:23:44	2016-06-28 15:20:53	2016-07-15 13:55:36
XQ008	老庄子镇中学	✔	2016-06-27 09:12:56	2016-06-28 15:20:41	2016-07-15 13:55:18
XQ009	宋各庄小学	✔	2016-06-27 18:38:19	2016-06-28 15:20:00	2016-07-15 13:55:41
XQ010	高庄子村	✔	2016-06-27 11:46:23	2016-06-28 15:20:40	2016-07-15 13:55:25
XQ011	崔家屯小学	✔	2016-06-28 09:33:20	2016-06-28 09:40:42	2016-07-15 13:55:45
XQ012	范家垴村	✔	2016-06-28 10:34:43	2016-06-28 15:20:31	2016-07-15 13:55:23
XO013	毛家垴小学	✔	2016-06-28 09:28:07	2016-06-28 09:41:16	2016-07-15 13:55:16

显示第 1 到第 14 条记录，总共 14 条记录 每页显示 50 ▲ 条记录

Fig. 6.28 Example of equipment monitoring data

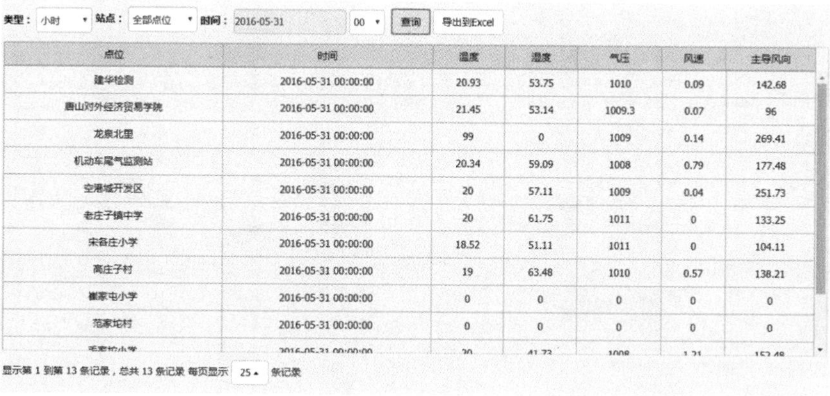

Fig. 6.29   Export of air quality and meteorological data

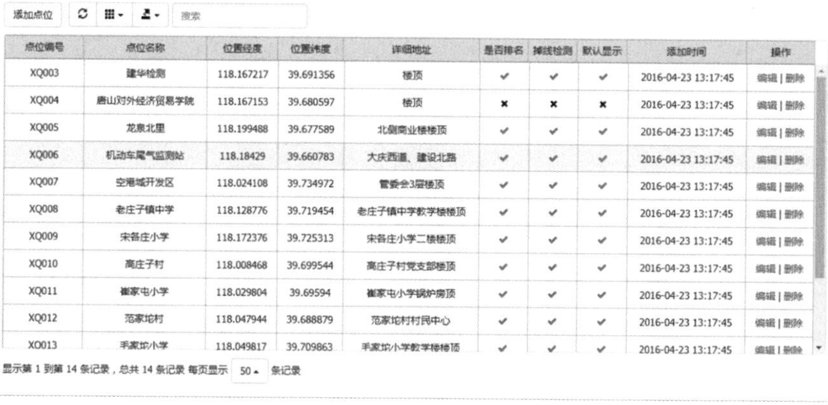

Fig. 6.30   Site selection

Fig. 6.31   SMS configuration

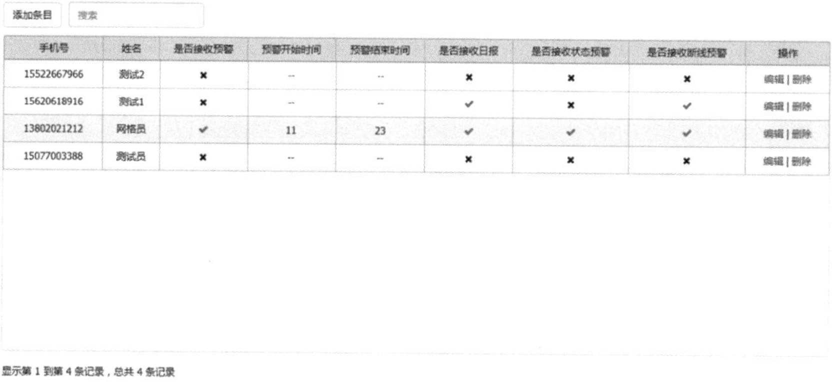

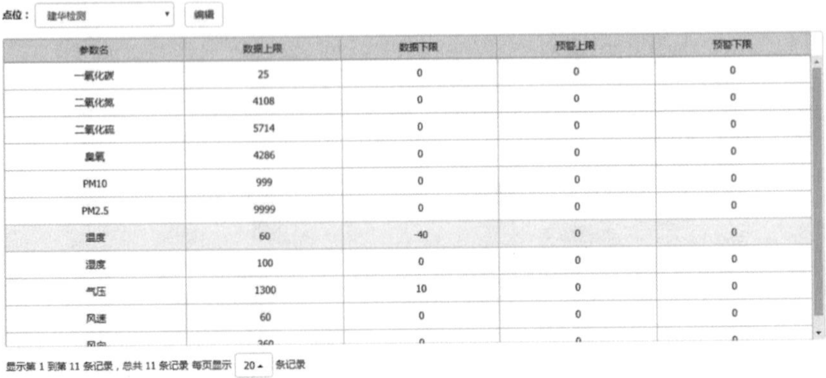

点位： 建华检测 ▼ 编辑

参数名	数据上限	数据下限	预警上限	预警下限
一氧化碳	25	0	0	0
二氧化氮	4108	0	0	0
二氧化硫	5714	0	0	0
臭氧	4286	0	0	0
PM10	999	0	0	0
PM2.5	9999	0	0	0
温度	60	-40	0	0
湿度	100	0	0	0
气压	1300	10	0	0
风速	60	0	0	0
风向	360	0	0	0

显示第 1 到第 11 条记录，总共 11 条记录 每页显示 20 ▲ 条记录

Fig. 6.32 Early warning of pollutant concentration

In case of abnormal fluctuation of air quality, the system will start the over standard alarm. This function is divided into the upper and lower limits of data and the upper and lower limits of early warning. The upper and lower limits of data are the standard values of data validity judgment. Those exceeding the limits are judged as invalid. When the monitoring factor is not within the upper and lower limits of early warning for a certain period of time, the early warning SMS will be sent (see Fig. 6.32).

20. Data rounding

The data rounding module can manually correct the wrong data not picked out in the program. The data rounding option can correct it. When the value is set to invalid, the data will be picked out and will not participate in the statistical operation, as shown in Fig. 6.33.

6.8.3 Application Function

1. Data analysis services

The data analysis service effectively combines the Internet data with the data of the standard monitoring station, analyzes and formulates daily, monthly, quarterly, semi annual, annual and special reports of air quality. The main local environmental problems are analyzed to define the direction of governance.

(1) The monthly report is required to analyze the air quality of this month. Compared with the air quality of the same period last year, the regional ranking is given. By analyzing the main problems of air pollution prevention and control in this month, we can obtain the control suggestions.

(2) The quarterly report requires that the air quality in this quarter should be analyzed for fine particulate matter traceability. Compared with the air quality

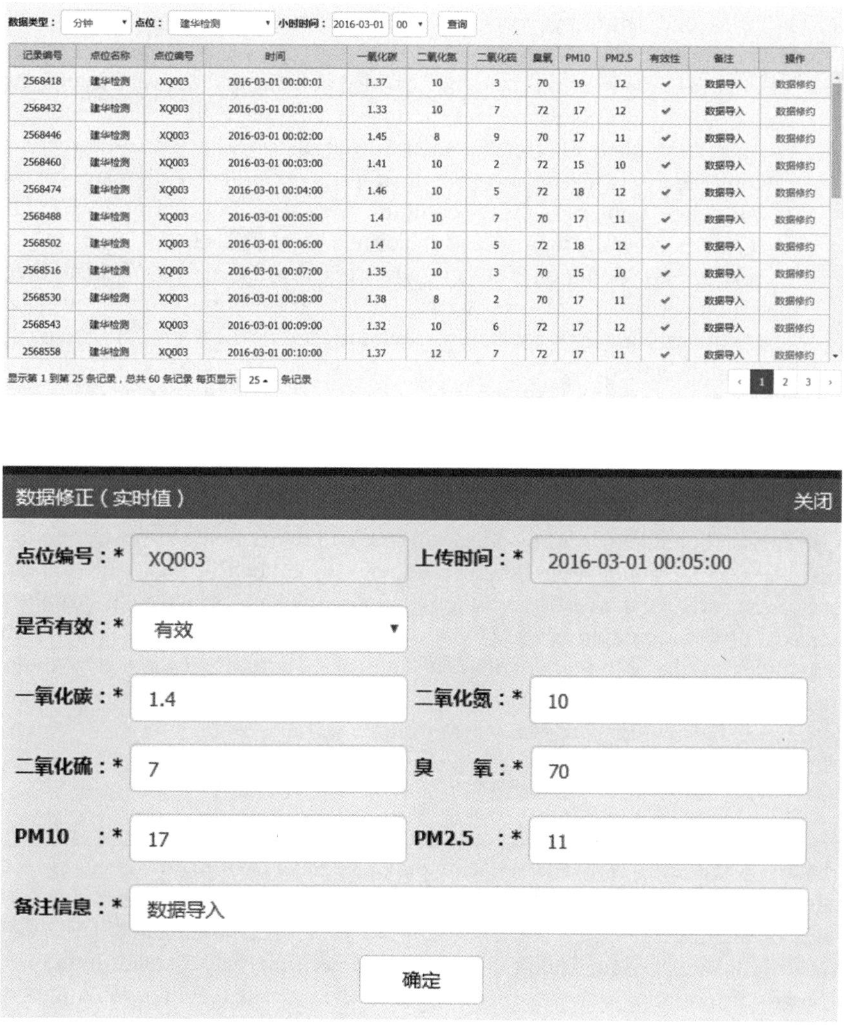

记录编号	点位名称	点位编号	时间	一氧化碳	二氧化氮	二氧化硫	臭氧	PM10	PM2.5	有效性	备注	操作
2568418	建华检测	XQ003	2016-03-01 00:00:01	1.37	10	3	70	19	12	✔	数据导入	数据修约
2568432	建华检测	XQ003	2016-03-01 00:01:00	1.33	10	7	72	17	12	✔	数据导入	数据修约
2568446	建华检测	XQ003	2016-03-01 00:02:00	1.45	8	9	70	17	11	✔	数据导入	数据修约
2568460	建华检测	XQ003	2016-03-01 00:03:00	1.41	10	2	72	15	10	✔	数据导入	数据修约
2568474	建华检测	XQ003	2016-03-01 00:04:00	1.46	10	5	72	18	12	✔	数据导入	数据修约
2568488	建华检测	XQ003	2016-03-01 00:05:00	1.4	10	7	70	17	11	✔	数据导入	数据修约
2568502	建华检测	XQ003	2016-03-01 00:06:00	1.4	10	5	72	18	12	✔	数据导入	数据修约
2568516	建华检测	XQ003	2016-03-01 00:07:00	1.35	10	3	70	15	10	✔	数据导入	数据修约
2568530	建华检测	XQ003	2016-03-01 00:08:00	1.38	8	2	70	17	11	✔	数据导入	数据修约
2568543	建华检测	XQ003	2016-03-01 00:09:00	1.32	10	6	72	17	12	✔	数据导入	数据修约
2568558	建华检测	XQ003	2016-03-01 00:10:00	1.37	12	7	72	17	11	✔	数据导入	数据修约

Fig. 6.33 Data rounding

in the same period of last year, the regional ranking is given. The main problems of air pollution prevention and control in this quarter should be analyzed, and the control suggestions should be put forward.

(3) The semi annual report is required to analyze the source of fine particles in the air quality in half a year. Compared with the air quality in the same period of last year, the regional ranking is given. The main problems of air pollution prevention and control in half a year is analyzed, and the treatment achievements and lessons are given. At the same time, according to the monitoring data of last year and the air pollution control in this year, the main problems

in the second half of the year were analyzed, and specific control suggestions were put forward.

(4) The annual report is required to be provided at the end of the year. It analyzes the air quality of this year, gives the source analysis results of fine particles, compares the air quality of the same period of last year, gives the regional ranking, analyzes the main problems of air pollution prevention and control in this year, summarizes the air pollution prevention and control work in this year, and points out the treatment achievements and lessons.

(5) Special report: pollution weather causes/traceability analysis. When severe pollution weather occurs in the area, a severe pollution weather cause/traceability analysis shall be submitted within 72 h after the end of the pollution weather.

(6) Special seminar: for key tasks, important activities and key projects, industry experts are invited to carry out special seminars.

2. Satellite remote sensing identification of contaminated areas

Based on satellite remote sensing data, using quantitative remote sensing technology, BDS and GIS technology, the satellite remote sensing identification pollution area service can realize the inversion monitoring of air pollutants, provide the spatial and temporal distribution map of PM2.5 concentration in various spatial and temporal scales, identify the pollution distribution status of the project area and surrounding areas, and issue the distribution map every day. The spatial resolution is 3 km and the temporal resolution is the daily mean value.

3. Daily air pollution source inspection

Daily air pollution source inspection service helps to establish an inspection team, which is 5 km away from the standard station. The key pollution sources such as industrial enterprises, construction sites, dust, motor vehicles, cooking fumes, all kinds of burning and open-air barbecue in the area were investigated, confirmed, reported and supervised. The problems found were timely distributed in the work group of air pollution prevention and control. The responsible units were urged to rectify in time and follow up continuously.

4. Forecast and early warning

The air quality forecast system is developed to accurately forecast the air quality in the next 72 h and provide the trend forecast for the next 5 days. According to the forecast results, the expert group puts forward countermeasures in advance to prevent. This will effectively reduce the harm and loss caused by polluted weather. Through the forecast results of this platform, the management department can timely, accurately and comprehensively know the air quality information and the development trend of air pollution. The platform provides important technical support for the implementation of air pollution joint prevention and control policy and the consultation work.

6.9 Application of Automatic Driving

6.9.1 Introduction

At present, the automatic driving agricultural machinery based on BDS has realized mass production and large-scale application. Automatic driving agricultural machinery has been introduced in detail in some sections of this book. This section mainly introduces the automatic driving technology in the field of passenger cars.

Autonomous vehicle, also known as driverless vehicle, is an intelligent vehicle controlled by computer system. Autonomous vehicles rely on the coordination and cooperation of artificial intelligence, visual computing, radar, inertial navigation, monitoring devices and GNSS. The computer can operate the motor vehicles automatically and safely without any human active operation.

At the beginning of the twenty-first century, automatic driving has shown a trend of approaching practicality. For example, Google automatic driving vehicle obtained the first automatic driving license in the United States in May 2012. In China, on December 18, 2017, Beijing launched the first domestic automatic driving standard. Beijing Municipal Transportation Commission, together with the Municipal Public Security and Transportation Administration Bureau and the Municipal Economic and Information Technology Commission, formulated and issued two rules: the guidance on accelerating the road test of automatic driving vehicles in Beijing and the Beijing road test management system for automatic driving vehicles. According to the two guiding documents of the implementation rules, it is clear that independent legal entities registered in China can apply for temporary driving on the road due to scientific research and type test related to automatic driving.

The two guiding documents are to locate the autopilot vehicle as a vehicle equipped with automatic driving system in accordance with the technical conditions of vehicle operation safety (GB 7258). Autopilot does not require the driver to perform physical driving operation. The automatic driving system can guide and make decisions on the driving tasks of the vehicle and replace the driver to control the vehicle to complete the driving. Automatic driving includes automatic movement, automatic transmission, automatic braking, automatic monitoring of surrounding environment, automatic lane changing, automatic steering, automatic signal reminding, network connected auxiliary driving, etc.

In the field of domestic automatic driving vehicles, Baidu driverless vehicles are relatively leading. In December 2015, Baidu first realized full automatic driving on the mixed roads of cities, ring roads and high-speed roads in China. On April 19, 2017, it provided an open Apollo software platform to partners in the automotive industry and the field of automatic driving, helping partners to quickly build a set of their own complete automatic driving system by the vehicle and sensors. The following text takes Baidu unmanned vehicle as an example to introduce the application of automatic driving technology and satellite navigation in it.

6.9.2   Key Technologies

Baidu automatic driving technology mainly includes four modules: high-precision map module, positioning module, perception module, intelligent decision-making and control module. It uses the high-precision map collected and produced to record complete three-dimensional road information, uses the combination of GNSS and inertial navigation system to achieve vehicle positioning in centimeter level accuracy, and uses traffic scene object recognition technology and environmental perception. The technology realizes high-precision vehicle detection and recognition, tracking, distance and speed estimation, road segmentation, lane detection, and provides the basis for intelligent decision-making of automatic driving.

1.   High precision map module

High precision map module widely applies deep learning and artificial intelligence technology to map production, which makes the unmanned vehicle have high-precision map data, and develops a high-precision map data management service system, which encapsulates the organization and management mechanism of map data, shields the underlying data details, and provides a unified data query interface for the application layer module, including element retrieval and spatial retrieval. It provides high-precision map solutions for unmanned vehicles with core capabilities such as search, format adaptation and cache management.

2.   Positioning module

The positioning module is mainly based on the GNSS and IMU. Combined with high-precision maps and a variety of sensor data, the positioning module can provide centimeter level integrated positioning solutions.

3.   Perception module

The perception module obtains the environment data around the vehicle through various sensors installed on the vehicle, such as LiDAR, camera and millimeter wave radar. Using multi-sensor fusion technology, the vehicle sensing algorithm can calculate the location, type, speed and orientation of the traffic participants in the environment in real time.

The perception module is supported by big data and deep learning technology. Massive real road test data are labeled by professionals and become learning samples that can be understood by machines.

4.   Intelligent decision and control module

The intelligent decision-making module can make comprehensive prediction, decision-making and planning of the unmanned vehicle. The module makes corresponding trajectory prediction and intelligent planning according to the real-time road conditions, road speed limit and other conditions. At the same time, it can give consideration to the safety and comfort, and improve the driving efficiency.

The control module can make the control and chassis interaction system of unmanned vehicle have accuracy, universality and adaptability. It can adapt to different road conditions, different speeds, different models and chassis interaction protocols. Its tracking automatic driving ability can make the control accuracy reach 10 cm level.

Chapter 7
Future Prospects of BDS and IOT Integration Application

7.1 From BDS System Innovation to BDS Application Innovation

BDS is a global satellite navigation system built and operated independently by China to meet the needs of national security and economic and social development. It is an important national space infrastructure to provide all-weather, all day, high-precision positioning, navigation and timing services for global users. China attaches great importance to the construction and development of the BDS. Since the 1980s, it began to explore the development path of the satellite navigation system suitable for China's national conditions. BDS formed a "three-step" development strategy: by the end of 2000, the BDS-1 will be built to provide services to China; by the end of 2012, the BDS-2 will be built to provide services to the Asia Pacific region; by 2020, the BDS-3 will be built to provide global services. By 2035, a more ubiquitous, more integrated and more intelligent integrated spatiotemporal system will be built and improved with the BDS as the core.

BDS is a model of comprehensive innovation of Chinese important projects. Starting from its own scientific and technological innovation, BDS project has led the innovation of enterprises and industries. From the perspective of technology, we can divide innovation into independent innovation, collaborative innovation and open innovation. The development of BDS just covers and reflects all the above innovations.

Independent innovation represents the original innovation in the process of technology research and development. There are many technologies in the autonomous and controllable BDS, which are the original innovations of China. From the short message of BDS-1 to the inter satellite link of BDS-3, these are the typical representatives of frontier innovation.

Collaborative innovation represents the use of the power of the team to achieve a breakthrough in key technologies. BDS has achieved key technical breakthroughs in satellite orbit design, satellite borne atomic clock, rocket launch, ground operation

© Publishing House of Electronics Industry 2022
B. Wang et al., *Internet of Things and BDS Application*,
https://doi.org/10.1007/978-981-16-9194-2_7

control, satellite measurement and control, etc. The smooth construction of each subsystem also ensures the steady development of the project as a whole.

Open innovation represents the enabling of technology to industry and the cross-border integration of multiple systems. BDS provides convenient and accurate spatiotemporal information for industry applications. The concept of "BDS + IOT" has penetrated into various industries. The industry applications have expanded from traditional transportation and logistics, emergency rescue and other industries to municipal pipe network, digital construction and other emerging fields. At the same time, it represents the sustainable development of technology. It is through continuous innovation that BDS can make continuous progress in accordance with the original "three-step" strategic plan and become an internationally recognized GNSS. BDS will become the core of the future national integrated PNT system.

BDS has advantages in constellation design, signal system and short message communication. It constructs faster than GPS, GLONASS and Galileo. With the rapid and steady development of BDS, in the face of the huge domestic application market, innovative applications of BDS also emerge in endlessly, such as driving test, driving training, pipe network inspection, etc.

The innovative application of BDS comes from the traction of demand, and also from the mining of industry pain points. For example, in the driver training and examination, in view of the problems of driver training equipment is not unified, information is not shared, examination efficiency is low, repeated investment is large, examination is not transparent, and equipment has many faults, a new generation of driver training system is built. BDS high-precision positioning and orientation technology is the core, which can realize the digitization and visualization of the whole process of driver testing and training, and significantly improve the quality of efficiency and scientificity of driver test and training. For another example, in the application of gas pipeline network, on the basis of BDS precise location service, the geographic information system, Internet, IOT, big data and other technologies are integrated with the gas pipeline network business. By using BDS precise spatiotemporal data, the construction management, intelligent inspection, anti-corrosion detection, leakage detection, emergency excavation and other precise management and control are realized in the process of pipeline network construction and operation. This can find the hidden danger in the pipe network equipment and management process, and to prevent the occurrence of pipe network safety accidents. Starting from the user needs, from the innovative design that brings value to the user, in the application environment, through in-depth analysis of the business process, the user participates in the whole process of putting forward the design scheme, technology research and development, verification and application. The user's actual and potential needs can be found and solved. The technological innovation of BDS will be promoted through various innovative technology and product applications.

7.2   The Coordinated Development of Spatiotemporal Information and the Internet of Everything

From BDS-1, a pilot system built in 2003, to BDS-2 serving the Asia Pacific region in 2012, to BDS-3 covering the world in 2020, the development process of BDS can be well combined with the global technology development trend.

From the end of the twentieth century to the beginning of the twenty-first century, the Internet wave penetrated everyone. The network has become an indispensable part of social production services, and the mobile communication business has also begun to enter thousands of households. During this period, BDS-1 was built and began to serve. In addition to military use, BDS is widely used in the field of fishery, especially short message service, which greatly reduces the communication cost of fishermen. The combination of mobile communication, network monitoring platform and industry application of BDS makes the civil application of BDS start.

From the first decade to the second decade of the twenty-first century is the period of rapid development of mobile Internet. With the development of 2G communication to 3G, 4G and 5G, smart phones and all kinds of apps emerge in endlessly. The convenient wireless connection mode promotes all kinds of intelligent terminals and background monitoring platforms to be widely used in the construction of smart city. At this time, the BDS-2 began to cover the Asia Pacific region. Different from the active navigation, positioning and timing service mode of BDS-1, the passive navigation, positioning and timing service mode of BDS-2 is consistent with GPS, which can be more widely used in the civil field. With the maturity of BDS-2, the prices of BDS/GNSS multi-mode chips, modules and terminals continue to drop. The prices of multi-mode chips have reached the same level or even lower than those of single GPS chips. The cost advantage plays a crucial role in the promotion of BDS civil applications. With more accurate spatiotemporal information brought by multi-mode enhancement systems such as space based and ground based augmentation systems, BDS industrial applications and regional applications are also deepening.

In this period, the IOT has also begun to develop rapidly. Various types of sensors, convenient wireless transmission, and analysis platform based on cloud computing big data have promoted the development of the IOT towards the direction of the Internet of everything. The IOT combined with BDS spatiotemporal information enables all kinds of business information to be perceived, transmitted, analyzed and served in a unified spatiotemporal framework. It should be noted that spatiotemporal information is only the starting point. How to combine it with business attribute information in different applications and provide decision-making by the analysis platform are more important, which need collaborative development. BDS needs to be combined with the IOT, and the implementation of spatiotemporal big data in various industries is the real collaborative development of spatiotemporal information and Internet of everything.

On the middle of 2020, BDS-3 has been completed, which can provide effective positioning, navigation and timing service for the world. Spatiotemporal information is the rigid demand of intelligent perception and the core technology of future intelligent service development. Integration and innovation is the inevitable choice of industrial development. Satellite navigation and location based service industry in China is moving towards a new stage of technology integration and industrial integration. The integration of BDS technology with advanced technologies such as communication, indoor positioning, automotive electronics, artificial intelligence, mobile Internet, IOT, geographic information, remote sensing and big data will present a new form of innovative development through the integrated application of terminal products and system services. Each link of the industrial chain will be integrated with high-end manufacturing industry, advanced software industry, comprehensive data industry and modern service industry to form a new business form of collective development. And this new form of technology integration and industry integration development and new business form can realize the orderly flow of people, money and materials under the precise spatiotemporal relationship, and ultimately bring people a real sense of "good in the sky, good on the ground" service experience.

7.3 BDS Spatiotemporal Intelligent Connection from the Perspective of Artificial Intelligence

The application of artificial intelligence (AI) includes accurate analysis and accurate execution operation. The two are complement each other and promote the development of artificial intelligence. The operation needs to be completed in a precise spatiotemporal framework. The BDS, which can provide accurate time and spatial location information for various applications, is the basis of the operation. Accurate analysis and decision-making depends on the quality of basic data. High quality data cannot only accurately reflect the characteristics of objects, but also contain traceable spatiotemporal information. Therefore, BDS is a powerful means to solve practical engineering application problems.

BDS can provide accurate and unified spatiotemporal benchmark, and the corresponding perception information is transmitted to the management platform through 5G communication. Artificial intelligence method is used for decision-making and analysis. The specific actuator forms a closed loop. The deep integration of the technology community composed of BDS, artificial intelligence and 5G communication with other industries can generate more industrial applications. It can not only empower and expand traditional industries, but also assist and enhance emerging industries.

Taking the traditional urban management industry as an example, the digital twin city will become a strong support to solve all kinds of urban management problems. The integrated PNT (positioning, navigation, timing) system with BDS as the core will provide a unified spatiotemporal benchmark. The low delay, low power consumption, high concurrency communication technology with 5G as the core will provide real-time, interconnected sensor networks. Deep learning with AI as the core and accurate execution will provide the improvement from automatic to independent. Therefore, from the perspective of artificial intelligence, digital twin city applications can be regarded as a collection of classification, recognition, detection and decision-making problems. Core technologies such as ubiquitous precision positioning, self-driven data acquisition, and knowledge-based big data service play a role in feature extraction. Infrastructure such as BDS precision service network, industry specific equipment and software platform, and data analysis platform play a role in feature aggregation. Precision positioning and timing, cross industry and whole process intelligent perception collaborative computing, and knowledge driven urban operation integrated decision-making play the role of matching features. They are finally implemented in the application scenarios of digital twin city, such as pipe network, power, water, sanitation, transportation, emergency, pension and so on.

With the continuous development of artificial intelligence technology, the application effect and scope of "BDS + AI" will be improved and expanded in all kinds of scenarios, from under driven with only a small amount of effective data, to independent event driven with all kinds of effective data, to data driven with a large amount of effective data, to knowledge driven with effective data.

Printed in the United States
by Baker & Taylor Publisher Services